One Hundred Years of Agricultural Education at Virginia Tech

One Hundred Years of Agricultural Education at Virginia Tech

Remembrances of Students, Faculty, and Administrators

Karen A. Vines and John Hillison (Eds.)

Department of Agricultural, Leadership, and Community Education
in association with VT Publishing
Blacksburg, Virginia

First published 2018 by Virginia Tech Department of Agricultural, Leadership, and Community Education in association with VT Publishing

Virginia Tech Department of Agricultural, Leadership,
and Community Education
214 Litton-Reaves Hall, Mail Code: 0343
Virginia Tech, Blacksburg, Virginia 24061

VT Publishing
University Libraries at Virginia Tech
560 Drillfield Drive
Blacksburg, VA 24061

Series enumeration supplied by publisher.
ISBN: 978-1-949373-04-2 (PDF)
ISBN: 978-1-949373-05-9 (epub)
ISBN: 978-1-949373-03-5 (paperback)

Every effort has been made to contact and acknowledge copyright owners, but the editors would be pleased to have any errors or omissions brought to their attention, so that corrections may be published in future editions.

Book cover by Karen A. Vines and John Hillison in consultation with Robert Browder.

Cover images courtesy of Historical Photograph Collection, Special Collections, University Libraries, Virginia Polytechnic Institute and State University University (top left, bottom right) and the Future Farmers of America (top right, bottom left).

Contents

Introduction

In 2017 we set out on a journey to learn about the history of the Department of Agricultural, Leadership, and Community Education (ALCE), which was then gearing up to celebrate one hundred years of agricultural education at Virginia Tech in 2018. Rather than produce a single narrative account, we decided to explore the department's history through interviews with former and present students, faculty, and administrators. The fourteen interviews presented in this volume are the fruits of that effort.

Throughout the course of our research we've had the opportunity to connect with some wonderful people who have been associated with the department over the past 70 years. One thing we learned early on was that this project could easily be an ongoing one with new collections of interviews published regularly. Not only did those we interviewed provide great insights, they also recommended others for us to interview in the future. Delbert O'Meara (M.S., 1967) even suggested the possibility of a separate volume focused specifically on Cooperative Extension. So, stay tuned; we may be asking you to participate in a future volume!

One particular insight that comes forth clearly in this volume is the intertwined history of agricultural education and Cooperative Extension at Virginia Tech. Dale Oliver (B.S., 1955) talks about serving as an Extension agent in Roanoke County on either side of World War II and then being called to campus to work on different projects, eventually serving as the director of the Extension option in Agriculture Education in 1986. Prior to this time, the primary focus of the department is reported to have been agricultural education, with some non-formal extension education included (Barbara Kirby, EdD, 1984). Delbert O'Meara (M.S., 1967) and others like him completed their degree working directly under Bill Skelton, Director of Extension, at the time. This predates the awarding of M.S. degrees by the department, which began in 1969 while the department was in the College of Education and provided a career path for Extension professionals who needed these degrees (John Crunkilton, FAC, 1969).

One of the most common themes in the interviews is that of movement of and within the department. The initial building in which the department was housed, the Agriculture Education building, is featured on the back of this book, along with Litton-Reaves Hall, which is the present home of the department. The department has been located in this Litton-Reaves since 1992 (John Hillison, FAC. 1976). Faculty and graduate students also spoke of regular changes in office location within the department. However, the most significant move as identified by several of the interviewees was the movement of the department from the College (now School) of Education to its original home in the College of Agriculture and Life Sciences. Many credit this move with setting the department on a course for growth into what it has become today.

Also, throughout the stories, the broadening of the mission of the department becomes evident. The initial purpose of the department, going back to when it was established in 1918, was "to produce teachers and provide in-service to keep teachers up to date" (John Hillison, FAC, 1976, p. 32). This led the department to develop the Agriculture Education Society and the Future Farmers of America, now known as FFA, Leadership Annual Conference, still hosted each year on the Blacksburg campus (John Hillison, FAC, 1976). In 1983, John Crunkilton (FAC, 1969) introduced an international component, working with the American Farm School in Greece. Other faculty and students have been involved from the establishment of the American Farm School through awarding masters degrees this year (John Crunkilton, FAC, 1969). Regina Smick-Atisano (EdD, 1988) reflects an expansion of the department by 1986 to increase degree offerings to include masters degrees for agriculture teachers and to prepare doctoral students to become teacher educators. Kerry Priest (PhD, 2012) describes the departmental role as a "very complex set of connected purposes that all centered around preparing people to do work that advances change" (p. 72). Much of this change is reflected in the changing of the name of the department in 2016 to Agricultural, Leadership, and Community Education.

A timeline is included to summarize the recollections of the interviewees and their involvement with events in the history of the department.

Methodology

This book is the result of fourteen interviews conducted by the research team, which consisted of three ALCE graduate students, Brittany Hoover, Asha Shayo, Lane Woodward and myself. Questions for semi-structured interviews were developed and revised based on feedback from my co-editor John Hillison and colleagues Tiffany Drape and Donna Westfall-Rudd. The study (IRB 17-710) was accepted by the Virginia Tech Institutional Research Board (IRB). The graduate students each conducted an initial interview as part of the pilot phase of the study. The questions were reviewed and determined to be effective in achieving the desired information. Brittany, Asha, and I completed the remaining interviews. Initial interviews took place in the fall of 2017 and the final interview was completed in September 2018.

Interviews were transcribed and reviewed for accuracy and clarification by the interviewees. Then I used the transcribed interview to create a narrative version, removing questions and adding clarification as needed to improve readability (Peters, Alter, & Schwartzbach, 2010). Interviewees once again reviewed in the narrative form and revisions were made. Then John and I made final edits for grammar and punctuation.

The book is published by ALCE in association with VT Publishing, which is a new initiative in the University Libraries at Virginia Tech under the direction of Peter Potter and Robert Browder.

Acknowledgments

First and foremost, thank you to those of you who participated in the interviews. We greatly appreciate you sharing your memories through this project and your commitment to working through the many revisions. Thanks to Asha and Brittany for the many interviews they conducted and transcribed. They invested a lot of time and energy into this project. Thanks to John Hillison for serving as co-editor for the book. And thank you for taking the time to read this book. I hope you will find it interesting and meaningful.

Karen A. Vines
Assistant Professor, Continuing Professional Education Specialist
Agricultural, Leadership, and Community Education Department
College of Agriculture and Life Sciences

October 2018

References:

Peters, S. J., Alter, T. R., & Schwartzbach, N. (2010). Democracy and higher education: Traditions and stories of civic engagement. East Lansing, MI: Michigan State University Press.

Timeline

This timeline was developed to situate the recollections of the interviewees within the context of agriculture and extension education over the past one hundred years. Included are both significant events for the interviewees as well as changes in the course of the department.

1914

Passage of the Smith-Lever Act provided federal funding to land grant universities for educational programs in rural counties for those who were unlikely to participate in formal education.

1917

Passage of Smith-Hughes Act provided federal funding to land grant universities for vocational agriculture programs provided through the schools in rural counties.

1918

The Agriculture Education and Animal Science Department began in the College of Agriculture. Purpose of the department: Produce teachers and provide in-service to keep teachers up to date.

1949

Establishment of the Vocational and Technical Education Division.

1950s

Delbert O'Meara began connection with Virginia Polytechnic Institute & State University as a 4-H member. Dale Oliver participated in the Corps of Cadets and completed his B.S. in agriculture education in 1955.

1960s

Strong federal legislation passed supporting agricultural education programs.

1967
Delbert O'Meara recieved M.S. degree in Adult Education and Extension, working with Bill Skelton in Cooperative Extension.

1969
John Crunkilton joined faculty. Program split between College of Arts and Sciences and College of Agriculture. Department offered B.S. and M.S. degrees through the College of Education. Al Krebs became department head, faculty interest in research increased and programming expanded in regional and national arenas.

1970s

Formed College of Education at Virginia Tech - teacher preparation programs brought together in College of Education from College of Arts and Science and College of Agriculture.

1971
Department moved to College of Education.

1973
Jim Clouse became Agriculture Education Program Leader.

1976
John Hillison joined the faculty.

1978
John Crunkilton became Agriculture Education Program Leader. Jay Poole graduated with B.S. in Agriculture Education; had choice of graduating from College of Education or CALS.

1980s

Established a scholarship program with $300 donation from the Agriculture Education Society (AES). Two to three female graduate students enrolled in the Agriculture Education department; no female faculty. Women first hired as teacher educators in ag ed programs.

1981
Department started the AES/FFA Leadership Conference.

1982
Barbara Kirby brought the first female FFA president, Ms. Jan Eberly, to Virginia Tech.

1983
John Crunkilton began work with the American Farm School in Greece. Carlton Everhart graduated from the departent with degree in agriculture education with an emphasis in agriculture business.

1984
People began thinking about proper administrative homes for agriculture education departments in the U.S. Program focused on teaching and learning in the schools and non-formal extension education. Program had a strong historical focus. Barbara Kirby worked as a graduate student with Bill Camp in integrating technology in the classroom. Barbara Kirby graduated from the department with an Ed.D. degree in Agriculture Education (teacher education), rural sociology cognate. Mike Rush graduated with Ed.D. degree in the Ag Ed Division.

1986
Purpose of the department was to prepare undergraduate and masters students to become agricultural teachers and prepare doctoral students to become teacher educators. Dale Oliver became director of new Extension Option in Agriculture Education. Regina Smick-Atisano was the second female student in doctoral program.

1988
Regina Smick-Atisano graduated with an Ed.D.

1990s

Extension program development and evaluation moved into the department.

1991
Memorandum of Agreement signed to move the department from the College of Education to CALS.

1992
The department moved into CALS and Litton-Reaves Hall.

1993
John Crunkilton served as director of the Agricultural Technology Program.

1994
Rick Rudd graduated with his Ph.D. in Career and Technical Education.

2000s

2006
Rick Rudd became department head of Agriculture and Extension Education; Department had 2 Ph.D. faculty and one tenure-track faculty member. B.S. in Agriculture Sciences approved by SCHEV.

2008
Doctoral program was fairly new. The Residential Leadership Community (RLC) merged with the AEE department.

2009
Chaney Mosley enrolled in third year that department had doctoral students on campus. Graduate students assisted with data collection in VCE listening sessions. Department set up facial recognition lab.

2010s

2012

Chaney Mosley completed Ph.D. in Agriculture and Extension Education. Kerry Priest graduated with Ph.D. from AEE with emphasis in collaborative community leadership and leadership education. Greater collaboration developed between academic and student affairs. There was also increased emphasis on service learning. Discussion about changing the name of the department from Agriculture & Extension Education to Agricultural, Leadership, and Community Education (ALCE). Department role seen as very complex set of connected purposes that all centered around preparing people to do work that advances change.

2014

Departmental name changed to Agricultural, Leadership, and Community Education.

2016

Rebekah Slabach received B.S. in Ag Sciences with leadership and social change minor. Department name changed to ALCE. The purpose of the department was to equip students to be leaders and engage with whatever organization they become involved with.

2018

Department had 18 faculty with mixture of tenure- and non-tenure faculty. American Farm School in Greece began offering a masters degree. Department celebrated 100th anniversary with gala at The Inn at Virginia Tech, on September 22nd, 2018. There were 125 people in attendance. This book was released at an associated Open House in conjunction with the CALS Tailgate on November 16-17, 2018.

1

Dale Oliver

B.S. 1955; Faculty 1960 – 1969; 1971 -1992

Professional Roles – Roanoke County Agricultural Extension Agent; Faculty training agricultural teachers in farm management; Research Assistant; Vocational Education project; Director of Extension option in Agricultural Education

I was first affiliated with agricultural education in 1951. That's a long time ago. I had completed a B.S. degree in agricultural education in 1955. I also was in the cadet corps and was commissioned as a 2nd Lieutenant in the U.S. Army. This presented some complications because I also received orders for active duty in the army in November of 1955. So, in a little over four months, I was to report to active duty. It was difficult to find a teaching job for a short period like that, so I was thinking about spending time on a home farm. However, I was contacted by a district Extension agent who wondered if I would be interested in a temporary position. There were agents in training at the time, so for the four months before I went on active duty in the army, I said I would be interested and I was interviewed and provided a position in Roanoke County.

So I was an Extension agent for about four months. Then I went on active duty in the army. Just before I completed my service of two years active duty, I was contacted by the district Extension agent. The position I had held temporarily was open and he wanted to know if I wanted to come back to it. I said yes, I would be interested, so I came back to Extension in November of 1957. I was an Extension agent for about three years. Then I received a call from the head of Agricultural Education at Virginia Tech. They had established a new position. It was to be a joint position between Agricultural Education and Agricultural Economics, and the purpose of the position was to provide in-service training in farm business management for agricultural teachers. I interviewed for the position and was hired with the stipulation that I would come back and do a masters degree in Agricultural Economics first, and then conduct conferences, workshops, provide curriculum materials, and do all I could to help upgrade the teaching of farm business management.

I pursued this endeavor from 1960-69. At that time, it was felt that the teaching had improved and the undergraduate preparation was improving too. And so, I had to decide what to do. I had taken courses in agricultural economics and economics, and I decided I would go ahead and finish the degree. And so, I was granted a graduate assistantship and I pursued the necessary research and completed a Ph.D. in Agricultural Economics.

After I finished that, I went back to Agricultural Education. At that time, money had been made available from the federal government to the states to conduct research and development activities. Since I had somewhat of a research orientation in my studies and agricultural economics, they asked me to develop a proposal for an activity, some type of a project. A colleague and I developed a proposal and it was funded. I was asked to direct this project. The project included myself – my salary was funded through the project plus three other faculty level research associates and some graduate assistants. And so, I began that project in 1971. My position was vacant in Ag Education, of course, since I was funded by the project. The salary was used to hire someone for a year and then that person left. So for the next 14 years they divided my old salary between research associates who were doctoral students and I continued on in a project. And so, this went on, let's say for about 15 years, and it was a major endeavor.

At that time, we had established a new Extension option in the program.

The work we did on the project developed curriculum materials for different occupations. We used what's called a task analysis. We did this as part of Virginia's obligation in a multi-state consortium. We would go out and interview people on the job in that occupation and develop a listing of what skills are needed to succeed in that occupation. We did these in agriculture, but also in home economics, business education and other fields, too. For example, we developed a task list for veterinarian assistants. Another one we did was for timber harvesters – identifying the skills you need to be in timber harvesting. And so forth. We had quite a list related to agriculture. In addition, we developed a management information system for vocational education in Virginia. Anyway, after continuing that project for about 15 years, I decided to go back into Agricultural Education in 1986. At that time, we had established a new Extension option in the program. So I went back and became the director of the Extension option in Agricultural Education. I also advised graduate and undergraduates and so forth, you know, a full blown faculty position. In the early '90s there was a move to try to reduce the faculty in Education and in other departments too, and they offered a buy-out. So, in 1992, I retired from Agricultural Education, as it was called then. I was in Lane Hall in the old program that was over there and the move to Litton-Reaves took place in '92. I was retiring, so I did not get involved in moving to Litton-Reaves. Since then, I have been retired and connected with the department in different ways, but it's been a long time.

One change was a move from one of the buildings, Smyth Hall, over to other buildings across campus, when we became part of the College of Education. But the effect was not all that great. We operated a little differently, but pretty much the same in the College of Education. The biggest move of significance was to the College of Agriculture and Life Sciences, which occurred just as I was retiring. That was a very significant move, because we were able to see at that time in the College of Education where we had a Division of Vocational Ed and Agriculture Education was a program area – just a small group, you know, that dealt with ag education. Coming over here, they were able to establish the department and expand the relationship with Cooperative Extension and have our own graduate programs and it's a big change. That's probably the most significant change during my experience with agricultural education.

The department moved to the College of Education in 1971. At that time, the purpose of the department was primarily focused on high school teachers, agricultural teachers, and professional workers in the fields of agriculture. But later, of course, the program was broadened to include Extension. But initially, during the early days, like when I was a student and so forth, the primary thrust was to prepare teachers – high school teachers. And at that time, we did not have the extensive system of community colleges which now exists, so we would not be preparing teachers for community colleges.

Before we moved, there was a problem with funding, staffing, and it was difficult to expand the programs and programs options, so I think that was the biggest problem. Having some of the things we desired, like our own graduate programs and so forth – we didn't have that. We were very limited to what was offered through the College of Ed. So that was somewhat of a limitation.

We taught courses in ag education, but we also taught courses in vocational education, which would include others like home economics and marketing education and so forth. So, we were at some times expected to teach courses that were taught across the board in the other program areas. That was somewhat of a limitation in expanding our program. Just by the nature of the organization.

We overcame this challenge by moving. Dr. Hillison was very involved in the move back to the College of Ag. See, we had been in the College of Ag years ago. He was involved in that, and he saw what was going to happen. When the College of Ed was downsized and almost abolished – we still have a School of Education – there were going to be cut backs in terms of faculty and we knew that the Agricultural Education program would be greatly reduced in terms of faculty and staff. So he wisely worked out the move to come back to the College of Ag. It was a very desirable thing. It just almost . . . you can list all the things we gained from the move back, and we probably would hardly exist today if we had not moved. Put it that simply.

(Y)ou can list all the things we gained from the move back, and we probably would hardly exist today if we had not moved. Put it that simply.

During the time, the department was pretty active in terms of national activity: conferences, publishing, and so forth – back in the old days. But I am sure that has been greatly expanded. For example, I was editor of a journal for two years and we published quite a bit and attended conferences and made presentations. We tried to maintain a strong professional visibility, but it was not the best situation – best organization to do what we wanted to do.

The department influenced my career in the sense that circumstances, like when I graduated I was commissioned in the army and called to active duty, and so it limited my options then and it caused me to go into Extension for a while and then come back into Ag Education. That was circumstances, really, but anyway, the opportunity came along for me to come back into Ag Education, so it's an interesting set of circumstances, probably unique. Being active in Extension and Ag Education, I think it was unusual.

I enjoyed working in Agricultural Extension. I had been there approaching three years when the opportunity came along and I thought that something new sounded interesting and exciting and it turned out to be that way. I didn't see the opportunities for advancement necessarily, at least not for some time and I worked in agriculture and 4-H. I did both of those things, so when the opportunity came along, I thought, "Well why not try?" It was a challenge. And of course, it involved getting into a different field – Agricultural Economics, and that in itself offered some challenges because it is somewhat of a technical field and I had to do some make-up study too in terms of mathematics and statistics and so forth to get through the program.

When I had the project, it was a 15-year project, I provided employment for 3 faculty research associates, but I hired and provided assistantships for a number of graduate students, I'd have to sit down and count them, but it was quite a number, so I think it was beneficial to them because they were fully funded in a graduate assistantship.

At the community level, I was involved most directly in the Town of Blacksburg. I was a member of the Blacksburg planning commission. I do not know if you are familiar with that, but they're advisory to town council on planning, zoning, and things like that. I was on the commission for 35 years and chaired the commission for 7 years at least. It was interesting because you knew what was going on in the community, development and so forth. For the last number of years our monthly meetings were televised, so that was a new experience – being on television. It was on local cable, and so I guess that was one of the biggest community involvements I had.

Since I retired, my wife and I funded an endowment to provide scholarships for students. That's one thing. We attend various activities. I come to meetings. The retirees from Agricultural Education get together about once a month and share stories and so forth. We have breakfast meetings and so forth and those are examples of my envolvement with the department. And serving on the anniversary committee is another example. So, I have some involvement.

What am I involved in right now? We are involved in church activities, for example. Let's see, I was on the board for a college for several years. We have my wife's home place, which is located about 90 miles from here, and we mow the yard and take care of things there and that is a two-day deal, usually, to drive down and back and mow the yard and things, so that takes time. Those are just examples... that somehow or other I stay busy.

I think funding will be a continual challenge for Agriculture and Extension Education. I don't know if that will ever disappear. I think the new leadership program that the department is in now is a challenge and maybe a great opportunity. I think there are a lot of opportunities in international programing and also in providing international students coming to be a part of our graduate program is a great opportunity. There may be others – I am pleased to see the relationship with Cooperative Extension become strong again. In fact, years ago, it hardly existed as such, and I think combining and bringing the two together in the department has been a really strong move. I see that as being profitable in the future. I guess that is pretty much it.

There are a lot of challenges for a new faculty member. There is a lot of an expectation for publishing, you know, and that's always going to be a problem, I guess. Securing grants and contracts and so forth to fund projects is a challenge and probably always will be. I guess that's pretty much it. I guess another one, somewhat of a challenge is getting and keeping the enrollment up, particularly undergraduate enrollment, and graduate too, but particularly undergraduate enrollment is a critical area.

Greatest opportunities, I guess are to just expand what you are doing. I think the direction is good. It's the opportunity to expand and further develop into what they are doing, I guess I would see it that way. As far as adding new things, I think the leadership option was a good thing that happened. It was a very interesting story about it because the provost at the time said, "Well would students want to enroll in that if it was in the College of Agriculture", and someone said, "Why don't you ask them?" Well, they did ask the students, and they didn't care where it was as long as they got the program, the option. They didn't care if it was in the College of Agriculture or somewhere else. I think it was properly placed here because we have people with leadership experience and professional preparation in leadership, so that was a profitable move, I think.

I've shared most of my thoughts, and I say keep going and keep building the program, because I think it's a great opportunity, and I am on an e-mail list about opening and opportunities in Extension, and I was surprised at how many opportunities there are across the country in Extension. Because it went through a period of being down as far as funding and so forth, and so it seems to be making a comeback and I think that is good. I would be very supportive of that.

If you can give one person credit, Dr. Hillison engineered the move from education to agriculture, so you'll find out about that, I am sure. He'll have a lot to say about that, no doubt.

2

Delbert O'Meara

M.S. CRD 1967

Professional Roles – Extension Agent; Role in Southeast District Office for Virginia Cooperative Extension

My first experience with Virginia Tech – which is now Virginia Tech, at that time it was Virginia Polytechnic Institute, was as a 4-H club member along in the early '50s. As a 4-H member, I was probably 9, 10 years old. And one of the first things I can remember that was really significant, and it was – you have to understand what the situation was. It was right after World War II. We did not have a lot of agriculture technology like we do today and a lot of things. And one of the things probably that was lagging in the whole country was purebred livestock, things that made livestock grow and become profitable. And Sears Roebuck – which wound up being Sears as a department store – but at one time, Sears Roebuck had a tremendous sales catalog with agricultural materials. You could buy a David Bradley mower, a David Bradley tractor. That was their brand name, David Bradley, which is now a collector's item. But they sponsored a thing called the Sears and Roebuck Pig Chain. And everybody says, "What in the world is that?" Well, I'll just explain it a little bit, because it helps to understand. We were all in rural counties. Even Northern Virginia, Prince William, which is now just as heavily populated as Fairfax. We had 300 dairy farms in Prince William County. And I lived in Prince William, right in the middle of dairy farms, and I was a 4-H member and we got five gilts and a boar given to the county as a 4-H pig chain. And the deal was, you got a pig, you got a gilt, and bred her and you raised pigs, and you had to give a pig back in the chain and it built in the county. Or we'd pick up these five, six pigs and give them to another county totally somewhere else. And this started out with about four or five pig chain counties in the state. And over a period of five or six years, they multiplied the purebred swine business.

And that really rolled the thing up for the swine industry. Got a better start, faster growing. I remember very well being involved with a Hampshire swine breeding thing and getting their magazine. And their field men came all the way from Peoria, Illinois. They would visit us here in Virginia. Mr. George Herring was a swine specialist in animal science that put this together. And he retired and Roy Gotze took over as the swine specialist and he worked with us on that. Basically out of that, my brother and I and my dad got in the purebred hog business and were members of the Virginia purebred swine breeders. And it was very instrumental in what I have done the rest of my life, really. That was basically the big thing that got me interested in agriculture and Virginia Tech and the career that I pursued.

But we had a very little bit of any kind of thing in this country that went on in World War II except producing metal, and producing airplanes, and producing ships. From the late '30s on through to about '47, we were basically a war-producing country that was torn, just putting our resources in war. And I think that's important when you look back at the college and everything, because all the students basically came out of the college and went into military service. Now, when you talk to some of the people who have been around awhile, they all came back and we had G.I. bills and stuff, and the universities, including Virginia Tech, just was loaded with all these people that were getting their college education paid for through the Defense Department. And I wasn't ever involved. At that time, I was eight, nine years old. But I've got a lot of friends that finished Tech after World War II. And they had all this Radford arsenal that was war production. And they had barracks that people lived there and produced ammunition and stuff. They all left and the whole barracks thing over to war plant is where most of them were housed and they rode school buses back over here to class. That's an important part of the history of the College of Agriculture.

Tech set up a masters program in Adult Education and Extension.

I graduated in, I guess, '56, '57 in Manassas and I went to Ferrum Junior College for two years. And that was at the advice of Mr. George Litton. He was the department head of Animal Science. All of the freshmen – because at that time, the school was building the School of Engineering and High Technology. And if you came out of a little rural high school, you didn't have much of a possibility of getting through the math and all the stuff that the college had set up. And so he'd ship all the ag students to Ferrum and then back up here. And I only went to Tech actually two years. And from that point, I went to work at Nansemond County, which was Suffolk, in 1962.

Well, I worked in Suffolk two or three years, and then they decided that a lot of the young people needed to have masters degrees. And Tech set up a masters program in Adult Education and Extension. And Bill Skelton was one of the real pushers of that and he made it financially where you can come to school and still work part-time for things that they wanted to do. I've done special little projects for him someplace within the colleges as well and worked on my masters. And there was a lot of people that went through that program. It was set up – there was the department on its own. But basically, it was handled within the field extension administration. The man that headed it up, he was hired as the director of training for Extension and he put that thing together. His name was Heckle. Older man. Can't remember his first name; he was from New England and he went back to be Extension Director, I think, in Connecticut or somewhere. But he was a very nice fellow and he understood field operations.

"Your job is to produce kings. Not that you need to become one, but you produce many of them."

And one of the things, I think, that really has contributed to the field of extension, agricultural extension in Virginia, and back to the college, was World War II veterans that became district agents and leadership. Because most of them were World War II officers. Most of them had a rank of captain or above. And Bill Skelton sort of fell in as their leader. And they had an ability to get you tied in and how to work things and how you work with people. That was a great experience too – and I'll never forget when I went to work in Nansemond County, in South Hampton County, there was an old gentleman named Red Davis, and Edgar Rauls was in Surrey, and Johnny Freeman was in Suffolk, or at that time it was Nansemond. And they all, when they carried me to lunch, they would sit down and they had a very interesting way of going that would get you involved and train you and making you know what your job was other than just being a person who pushed that technology. Technology changed so fast and you had to adapt yourself to it. They knew how to work with people. And how to work things. And I always remember what Red said. He said, "Your job is to produce kings. Not that you need to become one, but you produce many of them."

And that was the way Extension worked. Today, I don't see that. Everybody comes to work, they stay two, three months, take another job, or they move on, or they live 80 miles or 100 miles away from where they're working. And they go home at 4 o'clock in the afternoon and they're not – and there's something that came later that fit very well with that. And we had a fellow by the name of Paxton

Marshall that was in the Ag Econ Department. And that department had a little section called policy. And Paxton had a saying. He says, "A person working on a program or working for anything, or developing anything, has to be in and of." And what he meant by that, you need to be involved in the thing and you had to be a part of the organization that's putting it together. You couldn't work in a county and live somewhere else and the only interest you had was to show up at the office that day. And he did a lot for helping Agriculture Extension Education out of the college and all over the state even though he was here for a short period of time.

I think one of the greatest things is that we've worked through a number of different ways that – when I went to work, there was very little bit of money, a very little bit of salary. And one of the things that the leadership that had come out of World War II understood and the leadership here on campus understood, we went through several different ways of getting finance and salaries. I don't mind telling you. Extension agents went to work for about half of what everybody else was making because they just didn't have the money. And the big thing was you had to produce programs with leadership that wanted to fund you. And it started from the board of supervisors in the county all the way up through the state general assembly and the governor. And the leadership we had had the ability to become a part of that vast thing and got ourselves developed within that. We were in that whole system. That's where we got to, where we grew and had money, and became a – I won't use the word, powerful, but we came to a very influential thing within state politics and local politics and getting people educated in particularly the agriculture field. That was just a part of the total big complex, too.

There were a number of challenges. And we worked through all of them in a very successful way, including integration of the total Extension thing with Virginia State. Okay. I don't remember the exact dates, but I remember very vividly how we'd combined the offices at the one time. We had regular black offices from Virginia State and Virginia Tech, or VPI, offices and the federal court system rule was that you couldn't have two. And I remember very vividly when we put those offices together and how there were some trying days. But how it finally worked out where it strengthened the whole thing. And it still strengthens.

I think one of the greatest accomplishments early in the Extension thing in my career was that we started to get a lot of good data and a lot of research, and it worked, and we had a whole blossoming of a new area of agriculture. And I guess the best way to put it is that we went from hoeing weeds to spreading chemicals. And we had companies that really, after World War II, went to work and produced chemicals that worked, and Extension really had a great role in putting on the demonstrations that – one of the things in Southeast Virginia, we grew peanuts. And nut grass just ate peanuts up. You had a little nut in the ground, and the more you chopped it, the more it would multiply. And peanuts and these little nuts would blend together good, and the chemical company came out with a chemical that would kill nut grass but wouldn't kill the peanuts. And I will never forget – they came around and gave every county about 10 one-gallon cans, and had a thing for all the agents of how you applied it. And at that time, we didn't have much spray equipment. A lot of it was put on with a mopper type thing that didn't last for a year or so, but we had good sprayers and it would kill. And we'd just got enough to do about 10 rows across the field, and I remember very early that nut grass just turned brown, and the peanuts grew like mad, and the rest of them got ate up with nut grass. And the next year, there was enough of this chemical for the whole peanut industry. But one year, everybody saw the results, and that's how we worked and developed programs, and got people involved, and that was the start of the strength of probably the greatest food-producing situation ever been in the world. And we're still going through that same thing, and it's just the technology just keeps growing. The concern, I guess, if I have any, is that the university is doing good research, going through things. But industry, when there's a need, within six months, they can have it on the market. It takes six years to get something in the university. And we still have a role, but we've got to keep changing and look at how we stay in the picture. We've got to work ourselves to stay in the picture. Nobody's going to keep us in the picture.

Yeah. I've learned a lot working through the masters program. Things that we had people worried about doing that didn't know what they were doing and why they were working. We had academic kind of people that was trying to redevelop the Extension service and we were coming through. And I started through, practical knowledge people that knew where we were going and how to get us there. I had one or two professors in that program that were basically really great, but the rest of them; it was all purely academic and they had no idea how to fit. But we didn't really have our good department put together except for what we could find. I took a lot of – not a lot, but the main courses I took was over in architecture and urban regional planning, which really gave us a good view of that. And then we had several courses in the business department that – the courses weren't that great but it was who was teaching it that made it great. And we had some very good courses. I had ag econ, on land economics, some things that normally wouldn't put together to look at moving us up in a different level. Particularly, I came out in Northern Virginia working in Southeast Virginia, and at that time, not only was agriculture exploding, development was exploding. And I understood from where I grew up, from where I went to work and with my masters program that agriculture needed to be a part of the development. Don't fight it because you've got a farm in the middle of Fairfax County, middle of Virginia Beach. You need to look at what it's worth, and cash in, and move. And I worked with a number of people doing things like that. Today, there's not a dairy in Fairfax County, and there's only one and there will always be because its on a flood plain, land the owner won't ever have to pay tax on in Prince William, and between the two counties there was over 500 dairies in 1950.

I always took interest in new people coming on board, visited with them, made them feel confident that if they needed something, to call me. And that came from the district agent, Mr. D. T. Rogers. You know Mr. D.T., he wasn't what you'd call a supervisor, he was a friend. And that's what I tried to be. And those friendships still go on and on and on. And if you have a major problem, friends get along better than people that are directors. And I've had some tough days with things that we had to handle, misunderstandings or two people wanted the same power. And I can just go into a number of them but that's not important. But the main thing was that it goes back to "in and of". If you're going to work with the young staff, you have to be "in and of" that staff. I always carried you to lunch. We had something important about, always made an appointment for 11 o'clock and we went to lunch.

I still try to go to Extension activities, be involved. I've sort of lost track of working with – for a long time after I retired, a new agent come on board somewhere, I'd go introduce myself, take them to lunch, and leave a message with them that if you need something, give me a call, but now they just roll over so fast. By the time I get around to doing it, they're gone. And I used to keep up with everybody at the experiment station at Holland. I don't know none of them over there now. I know two secretaries and everything else has changed but it's been since 1991. And as we roll along, I have gotten involved in a lot of other things too. I was in the catering business for about 20-some years during that period of time. And then I got out of that and I wasn't going to do nothing. I was in the antique furniture collecting business and I got out of that, and then had a big sale and sold all of my warehouses and everything and said I wasn't going to do nothing. I was just going to relax and enjoy life. And I'm busier now than I've ever been. And then on top of that, I've gotten involved in buying real estate and redoing houses and owning rental property and buying and selling stuff. And we try to go to Florida and spend three to four months in the wintertime because that cold weather works on me so bad. But, hey, I go to Florida and I get involved buying and selling down there, so it's quite an experience.

It all comes about how you put things together. And I thought several times about writing the thing that they wanted. They wanted a resume for this awards thing, which I still don't understand anything about. And then, thank goodness, you should put this thing over there for me to read last night and I sat down and I thought through of how it went all the way back and one of the things that I've done early in life when I was in that 4-H pig-chain. We had a Veterans of World War II Agriculture Club in Prince William County, and they started a county fair. And it was all done with volunteers. At that time, we were starting up the hog business, and basically, myself and one or two other people put together the whole hog thing they developed. They done the same thing for Fredericksburg, and got involved back then when I was 12 years old, with the State Fair. I was always involved with something at the State Fair of Virginia, 4-H programs. I remember working with state 4-H horse show that they started at the State Fair, all up until they moved the fair, and they got a whole lot of people involved. They just spent money like mad and they went bankrupt. After they went bankrupt, I have not been involved with them at all. My career has made a great life for me to enjoy every day.

The biggest challenge we have, is that agriculture is a worldwide activity and opportunity.

Well, I think one of the biggest opportunities and one of the biggest challenges is that we've got to get young people, particularly in agriculture. I understand why everybody talks about the family farm, but when I travel around, and I see family farms, it's different than what most people have as the definition of the family farm. They're big, they're growing bigger. The biggest challenge we have, is that agriculture is a worldwide activity and opportunity. And I live within 15 miles of Smithfield, Virginia. The first time I went to Smithfield, I was 12 years old, and carried five boar hogs and four gilts, and sold at a purebred hog sale. They cleaned out an area of the little packing plant. Matter of fact, we had to wait for them to get the kill floor cleaned from that morning. We had the hog sale, put them in there at 1 o'clock, had the hog sale at 5 o'clock, and the trucks were waiting to unload the kill floor hogs for the next morning. At that time, they were killing 300 hogs at the plant. Today, that same organization, although it's changed and grew and been through five, six different new plants, they're killing about 20,000 hogs a day. And it used to be owned by two local families, and today it is owned by a company in China. And everybody said that when Joe Lulu sold out to the Chinese, the whole thing would go to H. It's unbelievable what they've done, what they've done for the community, what they've done for farmers. They doubled the kill, they own plants all over the United States within Smithfield Foods. And that's a good example of what everything's doing. The organization that you knew as Birdsong Peanuts, still is Birdsong Peanuts. But they're biggest operations, basically Argentina, and some foreign countries. Even though we still have a lot of peanuts in the United States, the whole thing, we've got to look at the international thing. And if you look at what's happened to the forest industry, it about dried up, the paper mills closed up, and all the need for things. And now, all of a sudden, we've got a worldwide demand for wood products, like fluff that you make diapers out of, that they're making so much of in southeast Virginia that's packed in big container ships, and the fluff goes to China and it's made in diapers and shipped back to this country. Well, we're getting into politics, all the things that make international trade, it's here to stay. Used to be, you went to North Harbor, and there was a few ships sitting in Chesapeake Bay waiting to come in. They've tripled the amount of space and put ports at Norfolk and Newport News, and Hampton, as well as they've got a terminal in Richmond now that they're running on barges. And there's ships lined up as far as you can see, going out to the ocean, getting in and load, and they load the round thing, and there's a constant stream of tractor trailor trucks. And that's about the last three or four years, and so there's where you've got to go and you've got to look at – it's important that kids and families who all have their own eggs, and it amazes me how counties and cities have zoning ordinances and how much time they spend advertising, changing the zoning thing, most places now have – you can have five female chickens, but you can't have a rooster. Virginia Beach probably spent a huge amount of money just for the legal advertising to eliminate the rooster in a backyard chicken situation. And the real fear in this whole thing, we can have one chicken, poultry flu thing, go through and it will take it all out and you'll never see it again. I don't have time now but I have a whole interview of how I got within the pork business, how I got in the chicken fancy business, which was a great thing in my life, too.

3

Jim Clouse

Faculty 1973 - 1987

Professional Roles - Purdue Faculty; Virginia Tech: Agricultural Education Program Area Leader; Professor of Agricultural Education

I came to Virginia Tech in 1973 and was here until I retired in 1987. I came as a head of the program area of Agriculture Education which was in Vocational and Technical Education and a part of the College of Education at that time. And then later the college disbanded and the department became a part of Agriculture. But I was program area leader, which was the same as head of Ag Education, from 1973 to 1978. In 1978, John Crunkilton became program area leader.

There really were not many changes in the department at that time. It was a very stable department. Things were operating very well. It was a progressive, strong department. We had an excellent faculty. Particularly in the early years. There were some changes, obviously. But as you may or may not know, back in the late '60s and early '70s, there was some very strong federal legislation in vocational education. And because of that, we had a very strong ag department. We had a great group of teachers. All we were concerned with was training ag teachers at that time. About 60% of our ag graduates went into teaching the first year, 40% did not. Within five years, it'd only be about half of that 60% still in teaching. So students were going into Extension, they were going into seed and fertilizer sales and all kinds of other work. Back to the farm and so on. We had a strong graduate program because of this federal legislation. There was a lot of federal money for people working on masters and doctorates. They came here to get their doctorate from Idaho, Arizona, North Carolina, Pennsylvania and other states. So it was a very strong graduate program at that time.

The three major challenges for any time are personnel, budgets, and program. And of course, you always have that and you have to work for budgets and you have to work for personnel. And a big factor, of course, is going out and getting quality students. Of course, we worked with the state department at that time, which was very strong, too, and getting good students coming in. That was a period of time when a lot more young ladies were coming into the program. I guess in the late '50s there were almost none, really. And in the '60s then they gradually came in and I think they're probably more than half now. Young women, particularly in the teaching part of the option. I expect in Extension and others, too, from what I hear. It was that transition in time. The faculty at that time was still all male. And I don't know when the first female came into the faculty. That was much after I retired. So I don't know. But those were the things that we're really concerned about.

And we had a lot of travel that we did. Both in-state and out. The staff were very active in national programs. They were getting involved in international programs at that time, which I encouraged, frankly, and was involved in to some degree, too. John Crunkilton was involved with Greece. I was in Greece. A number of years I worked in East Africa and South America and other places. At that time it was just beginning. And then in the 70s and 80s it became more prevalent.

I think the ag program was probably the strongest of any of the vocational education programs that were involved in the division. And that was noticeable to people. Part of it because I think it had been strong here, both in the college and, as you know, FFA background and other things – all of them led back here to Walter Newman, Harry Sanders, Henry Groseclose and some of the other early leaders. And incidentally, Walter Newman and Harry were still alive and I knew them well during the time I was here. They were both retired, but were well known. I think there was a history of strength in agricultural education. And it showed. And obviously, people recognized it.

Dr. T. Marshall Hahn, President of Virginia Tech, was my next door neighbor here for years and a good friend. He and Rufus Beamer and others convinced me that this might be a good move and made it attractive enough that I came. I think one of the big factors was, I was able to work with a strong state supervisor of Ag Ed and six excellent area supervisors at that time. I'm not sure how many they have now. But we had a strong relationship with Julian Campbell and his six area supervisors. We worked very closely with them. And I think that really was a strength. And then in addition to that, we got involved more in the national and international activities.

I saw my job as encouraging faculty to: 1) do an excellent job of teaching, and 2) do an excellent job of in-service work with teachers. We did a lot of night classes, and summer classes, and workshops, and things like that. Those two things particularly. And I encouraged them to get involved in national and international work and to have a strong graduate program. We worked to be a leader, particularly in this eastern part of the country. A little background – I spent a sabbatical at NC State from Purdue in '68 or '69. And I visualized, and I think I was correct – Tech, at that time, was getting a lot of students from the northeast quadrant of the country, perhaps because it was less expensive to come here than to go to Penn State or Cornell or Rutgers. And so we really pushed getting out-of-state students. And then they went home, of course, and acted as good ambassadors for Tech. But I think if you have a strong – I'm going to call it a feeder cadre, a feeder program, you get better students and you get more students coming in, and hence, better people graduating.

But getting along with people is today far more important than getting along with pigs or cows or something else...

When I retired, I had a farm here, which I sold to my son 20 years ago, and then I became a wood turner. I turned wood and showed it at craft shows and things like that, Stepping Out, for example, and things as a hobby until about six years ago when my eyes started going bad. And so I got involved in other things. And I did some international work up until about 1995 or so. Albania, and Greece, and Italy, and East Africa. Crunkilton, Hillison and Dale Oliver, and Stan Burke, and I, and Martin McMillian used to get together once a month for breakfast. And we still do, but there's only Dale, and John, and John, and Stan, and I now. We usually had breakfast once a month and discussed things a little. Some good, some not so good. Politics, and sports, and everything else.

I guess the last time I was back in the department was to see the historical room. I was a member of the committee that established the National FFA Alumni and second president of the alumni and hold life membership number three in the National Alumni. I went back to that room one time and met with the National FFA President and visited with him for a while.

Programs, budgets, and personnel – are only three of the challenges facing agricultural education and education today. I think there's some real problems today with turning out quality people in a variety of programs that I understand you're trying to prepare students for. You do try to prepare for teachers, Extension – I'm going to call ag communicators, these are sales, people. I'm not sure what else.

What all are you trying to prepare for? And I hope that it has not bled strengths and efforts away from what I'm going to call the basic programs of teacher education and Extension education and good community service, because to me those are three fundamental parts, and training of good people in those three areas is extremely and will continue to be, extremely important. You train enough people to supply the need for teachers and Extension agents for example. It'll be interesting to see, because I see today trying to not be all things to all people, but to be a variety of things, which means they need to have an end goal of some kind that they have some potential of getting into. And they need to be competitive when they come out in terms of price and benefits and location and everything else. It used to be that people tended to stick pretty close to their home state, and many times within a 50-mile radius of home. That's changed now. They even can go anywhere and do anything.

I think one of the big opportunities, really, is to make sure we stay committed to a strong base of good leadership training, good communications training, basic science training, because agriculture's based on science, and if we stay with a good base, we'll be all right. But getting along with people is today far more important than getting along with pigs or cows or something else, and if you can't get along with people and get them through leadership activities to do what needs to be done in the way that it needs to be done, you're going to be in trouble. We can turn out all kinds of scientists and experimental people that do research and that's great, we need those, but we need somebody that can take that research and through people do something constructive with it for the betterment of mankind.

4

John R. Crunkilton

Faculty, 1969 - 2003

Professional Roles - College of Arts and Sciences and College of Education: assistant, associate, and professor, Program Area Leader, Associate Dean; Interim Dean College of Agriculture and Life Sciences: Director Agricultural Technology Program, Associate Dean, Assistant Director Academic Programs

I came to the department in August of 1969 after completing my doctorate at Cornell. I was hired as an assistant professor with the main responsibility to teach the undergraduate methods course. That was the major focus and this continued as a major focus throughout my professional career. I retired in 2003. In 1998, I was on a leave of absence from the department and worked in the dean's office in the College of Education as Associate Dean and later as Interim Dean until the new dean came. When the Agricultural Education Program Area moved to the College of Agriculture and Life Sciences, I chose to transfer my appointment to that college. While I was still listed as a faculty member in the department, I worked in the academic dean's office with the appointment of Associate Dean, Assistant Director of Academic Programs, and Director of Agricultural Technology. I was not directly involved in the department during this time. I still had my tenure in the department but major responsibilities were outside the department. I think it was in 1992. While in the College of Education, the department was called an Agricultural Education Program Area. It was not a department, because at that time all the vocational service areas like distributive education and home economic education were listed as program areas in the Division of Vocational Technical Education.

So, at that time, their emphasis basically was what you might call the more practical side of teacher education, teaching students to become teachers.

In 1969 when I came to Virginia Tech, the faculty were, let's call them seasoned faculty. They had been here a long time. Most of them had masters degrees and the major role of the department at that time was to prepare students as vocational agriculture teachers for teaching in secondary and middle schools in Virginia. So, at that time, their emphasis basically was what you might call the more practical side of teacher education, teaching students how to become teachers. The faculty members were not interested in a national reputation at that time. They did very little research, if any. They taught, supervised student teachers, worked with agricultural teachers in the state, and worked closely with the state staff in Ag Education. This was the atmosphere in 1969 and, of course, that was pretty much the atmosphere, I think, in many programs throughout the United States that were focusing entirely on preparing vocational agriculture teachers. We continued to have as a major focus of preparing agricultural teachers through the '70s. But when Dr. Al Krebs came as head in 1969, the department moved to include more research, and an interest in building national visibility rose. That was a significant change for the department during that time. As these seasoned faculty retired, we hired individuals who had an interest in research and aspired to become national leaders for our profession. I think a significant change during that time was one from focusing not only what was needed in Virginia, but one to include the needs on more of a regional and national scope. Probably another significant change during that time was that the program area moved from the College of Education to the College of Agriculture and Life Sciences.

In 1970, the College of Education was formed at Virginia Tech. The decision was made at that time that any program or academic unit that dealt with the preparation of teachers would be administratively housed in the College of Education. When I came in 1969, I was actually in the College of Arts and Sciences and agricultural education students were in the College of Agriculture. Then, in 1970, when the decision was made to form the College of Education, all programs that dealt with preparation of teachers were moved to College of Education. At that time, our students were moved from the College of Agriculture to the College of Education. The College of Education formed four divisions, of which one was the Division of Vocational Technical Education (VTE). Since agriculture was basically a vocational preparation program in the schools, the VTE Division consisted of agricultural education, home economics education, trade and industrial education, industrial arts education, distributive education, and business education. It was that way until mid to late 1980s and during the latter years, the university basically did not see a high priority for the College of Education. There were rumors that it was going to be disbanded and eventually it was. At that time, I was not in Agricultural Education, as I had been assigned to the associate dean's position in the dean's office, and the Agricultural Education faculty made the decision that it was best for the future of the program to move to the College of Agriculture and Life Sciences. Shortly after that, I do not know what the time frame was, the university decided to do away with the College of Education. Parts of that college still exist today as the School of Education, but there is no longer a College of Education.

In 1969, the department basically offered the bachelors and masters degrees when the College of Education was formed. Later the College of Education added the doctoral program and we could begin to award the doctoral degree. So, I think the other important milestone that would add to a significant change was the ability to offer doctoral degrees. And of course, that would also mean that faculty had to have a greater interest in research, writing, and publishing. This was a challenge to the department, but the whole university was also changing in this direction. Also, when I came in '69, the university was still pretty much a military type of school and Virginia-focused. Through the '70s and '80s, they started thinking the university ought to be grander and bigger. Student enrollment was rising, and the emphasis on grantsmanship was increasing. And with the start of the doctoral program, it was critical that more emphasis and encouragement be made to award doctoral degrees, conduct research, produce scholarly articles, and give presentations at national meetings, all of these things were becoming a high priority. Plus, this new atmosphere related directly to success of the faculty in the promotion and tenure process. If you want to be associate professor all your life, don't do anything but just teach. If you want to achieve the rank of professor, develop a national reputation, and assume some leadership opportunities. Then a faculty member must get engaged in national and/or international organizations and work.

A continual challenge for the College of Education, as well as the Division of VTE, was adequate funding. Vocational Education was not high on the funding list for a lot of people. Most of the College of Education faculty were more research focused and experienced in grantsmanship, and VTE was preparing students to be teachers for the secondary school level. That is not very exciting to a lot of people, and so it was always a constant battle for funding for our programs. I think over that time, the biggest challenges were just getting the funding that would allow the faculty to do what you thought you needed to do – and that was preparing vocational agriculture teachers. At that time, we were not interested in Extension agent preparation, leadership, and the things that the department includes today. As the department matured over the years and added doctoral advising and research projects to the program, the faculty were able to be promoted and tenured. I think the faculty we had for the most part was able to get promoted and tenured because they were doing those things that the college and university valued.

I think that when we added a national component to our Virginia focus, I thought we had a pretty good reputation. We attracted some excellent doctoral students into our program and we were viewed as a national leader along that line, while at the same time, we continued to maintain the key focus of preparing vocational agriculture teachers. We had a pretty strong State Department of Education during these years. There were six Area State Supervisors of Agricultural Education. The state was divided into the six geographical areas and each had a supervisor assigned to it. While that would not probably appear to be an accomplishment, being able to maintain that connection, that commitment, that professional dedication, and team work approach to the Vo-Ag program in that state – we were able to provide a service to the citizens of the state for quite a long time. And during this time, not much state effort was put on students taking vocational education in high school. A big emphasis in high school was that everyone should go to college, take foreign language, and enroll in science courses to prepare them for college. As long as we were able to keep the area supervisors, we were able to keep the focus that vocational education was serving a vital role in the educational system. This was an accomplishment that we achieved and maintained, while at the same time developing a national reputation for the doctoral program. This national reputation was important for faculty promotion and tenure.

This was the only university I interviewed for when finishing the doctoral degree. Basically, I have to admit, I probably came in with blinders on. I grew up on a farm in Ohio and had ag in high school. I taught ag in Ohio, and that was basically my perspective of the field. So, I would have to admit when I came here, I was totally focused on teaching undergraduate students. That's why they hired me. After working with Al Krebs and other national leaders in our field, they really influenced my career. Later, I was selected to be editor of the AATEA Journal and then I became involved in other national and international leadership positions. So yes, the department did have a profound influence on my career from that standpoint.

One of my classmates from Cornell got involved with the American Farm School in Greece, which opened up the opportunity to earn a Fulbright Grant to go to Greece. In fact, eventually I ended up with two Fulbrights to work with American Farm School. Being a faculty member in our department provided a rich experience which led me or gave me confidence to do the type of things I thought I could do to help with the American Farm School in Greece.

Then came the opportunity to work in the dean's office in the College of Education certainly. I was somewhat of an old-timer by that time. So, I had a lot of history about university governance, college governance, departmental governance, and state department relationships. So, I think all of those things, if you talk about how it influences your career, well, that gave you really a solid background on which to carry out your responsibilities at that time.

When I came over to the College of Agriculture and Life Sciences, I still had an appointment within the department, but I was housed in the Associate Dean's Academic Affairs Office. Later, I became director of the two-year Agricultural Technology Program, which really fit my background quite well because agricultural technology was basically a technical education program. A lot of people felt it should not even be at Virginia Tech, that it should be located at a community college. Some faculty felt it was below their dignity for this program to be in the college and yet with my vocational education background, to me, agricultural technology was a very nice fit. Where else could the students get the most or the latest technical information in agriculture than here, at the land-grant college in Virginia. Students could get some of this level of education at certain community colleges, but here at this campus is where the latest technology of agriculture was located, so what better place could there be? I had the philosophy and experience to see the value of this program in our total educational system. How does that relate to my career? I was director of the program for 10 years. I always thought that it was a good program. It provided a need for Virginia agriculture that was not being served. Sure, the community colleges had some agricultural programs, but they were always limited on facilities, experienced faculty, and student numbers, they could never generate enough numbers in some of the programs to justify offering the courses. There were a few community college programs upset with Virginia Tech for starting the Agricultural Technology Program, because they would see that as competition. Which it was, but the program here has been able to survive.

I worked closely with the American Farm School in Greece since 1983. This fall (2018), it started offering a masters degree.

I would like to think that you would have been able to have a positive influence on all your doctoral students and advisees you have had over the years. When I look at some of the doctoral students

I have had, they have taken pretty significant career positions. So, you like to think you might have done something to help them along the way. I think with the international work, I worked closely with the American Farm School in Greece since 1983. And I have helped to start a college at the Farm School. The college started out as a two-year program offering an associate degree, and later became a four-year offering a bachelors degree. This fall (2018) it started offering a masters degree. While it wasn't my idea to start the college, my second Fulbright Grant to Greece was basically to develop what's called the infrastructure for the college–for example, meaning how you determine the curriculum including courses to be offered, setting up timetables, establishing advising procedures, setting up student services, and developing a recruitment program – that is helping them to establish all the infrastructure that would be needed to get the college started. I hoped that has had some positive influences. After the first Fulbright Grant, the Farm School offered summer leadership programs in which they invited agricultural leaders from developing countries to attend. They ran this program for six years. We had maybe 20 participants each year, representing school principals, agricultural teachers, regional extension agents, and ministers of education. We brought them to the Farm School for a five-week course on both practical education and how you teach practical education. And for leadership – how you plan educational programs. After 5 years, we had close to 100–120 participants which we hoped would return to their countries to assume leadership positions in their educational programs. Many of these participants came from Africa, and I hoped we had a positive influence on their lives and careers.

My time here gave me the opportunity, it opened doors to me. Like I said earlier, when I came in '69, I was really green and inexperienced in the professional field of education. All I knew was that I was a farm kid lucky enough to go to college, earn a doctorate and come here and that was it. Teaching students to be good agricultural teachers, and that was a pretty narrow focus. I think through the years all these other opportunities and experiences fell into place.

When the department moved to the College of Agriculture and Life Sciences, there were very few full-time faculty. When Dr. Rudd came as Head, I was already retired. The department was too small to form a promotion and tenure committee, as Dr. Rudd was the only tenured professor. The university allowed Dr. Hillison and I, plus one faculty member outside of the department in the college, to serve as the Promotion and Tenure Committee. Today, the department has grown and enough of the faculty have been promoted and tenured to where Dr. Hillison and I are not needed for this important role. I think by working on this history committee, I have helped to develop some of the department's history that eventually might be used in the department's 100-year celebration.

Anytime somebody thinks that I might be able to add something to the department I would be open to working with them. I don't know what that might be, because the longer you are retired, the more you don't know about the department....its mission, its goals, its activities. Sure, I might know most of the faculty and have an idea of what they are doing. But the day to day, the things that need to be done and that are done, I really don't know what their thinking is, their visions, and so a person may not know exactly how you would fit in.

The only term I can think of is...re-branding...you know what branding is and re-branding, and I am not sure if that is a challenge or an opportunity for the department. I know the department now still has as one of their focus that of preparing individuals to be teachers of agriculture. We did a little bit of preparing extension agents when I was in the department, but this was not a major focus. But now that is one of the major focus in your department, as I understand. You also have your leadership component, and I do not know what else you may be doing. So, to me the challenge or opportunity is how or how do you re-brand the department? How do they see themselves wanting to be known within the university and nationally, and I don't have the answer to that question, but to me it would be a challenge no matter what. The other challenge is not losing focus of preparing agricultural education teachers. I know it is difficult today with the state and university not giving a lot of emphasis on teacher credentialing and it is out of your hands. So, the challenge to me would be if they still see that as one of their major missions is how are they going to try to operate in this environment? As I reflect back, it was easy when I was here to prepare these types of individuals because you had a state supervisor in Richmond – very strong, and that was how the State Department of Education operated. You had state supervisors for home economics, agricultural education, distributive education and so forth, who really championed the state preparing vocational education teachers. And then there were six area supervisors who worked very closely with the school districts, superintendents, directors of vocational education and championed quality programs. Reflecting back, even though we thought we had some challenges at that time, but as compared today, if you want to keep that as a major mission, it is a much more challenging problem for the department than it was when I was here. So, how do you continue working with the teachers when you don't have this type of organizational framework to work with? I don't know how do they do it. Or how should they do it? Or how could they do it better? Or what are the opportunities there?

My more traditional view point is, how does the department maintain this contact with the agricultural education teachers in the state? How do they maintain that contact? If they want to, maybe they don't want to, but how do they put in place the opportunity for the faculty to keep in touch with these agricultural teachers? I know this is a challenge with your leadership component, with the Extension, and I do not know what other components you have in the department now. But how do you mix all of those missions together and make sure each of them receives the structure for it to successfully occur?

When I came through my work here, and when you think about preparing agricultural teachers, we relied a lot upon the writings of Dewey, Lancelot, Hammonds, and other earlier writings in the profession on how teachers should teach, the pedagogy. What do the faculty today hang their hat on when they walk into the classroom to try to teach students? Are they or do they have a philosophical foundation or research findings on which they are basing their own teaching styles and practices? I don't know. I am not sure if they are using Dewey, Stewart, Lancelot or Kreb's problem solving approach. Or, when they talk with their students about educational psychology and how students learn. What is the new research in the field showing on how students learn? Is that being infused into the curriculum now, whether it be the teacher preparation, extension, or leadership component? I don't know, but to me, it will always be a challenge as to how the department and the individual faculty members keep up to date in their own profession.

5

John Hillison
Faculty, 1976 - 2006

Professional Roles - Program Area Leader, Director, Department Head

I served as a faculty member and an administrator with the department. My administrative titles were program area leader, director, and department head. I served from 1976 to 2006, when I retired.

The department had three names: first, it was the Agricultural Education Program Area within the Division of Vocational and Technical Education in the College of Education. Then when we moved to the College of Agriculture and Life Sciences, it became known as the Agricultural Education Program. Then when we merged with the Extension unit, it became Agricultural and Extension Education and that's when we formed the department.

Departmental history in numbers: 3 Colleges; 14 Deans; 13 Departmental Leaders; 88 Faculty

The most fundamental changes happened when we came back home to the College of Agriculture and Life Sciences. That is where we had started in 1918. That change occurred in 1991. In 1918, when the department was first started, it was known as Agricultural Education, and in 1949, that led to the establishment of the Vocational and Technical Education division. The history on that is Agricultural Education to Vocational and Technical Education and which, ultimately, the College of Education could be formed. Essentially, Ag Education started all of that. Then ultimately, we came back home to the College of Agriculture and Life Sciences. There was a formal memorandum of agreement signed by both the dean of the College of Education and the College of Agriculture and Life Sciences on December 4, 1991, to move the department into the College of Ag. The physical move to Litton-Reaves occurred on July 1, 1992. With those changes, we lost our departmental status. We lost our budget. We lost our degrees. We lost our students, and the student majors. We sacrificed all of those so that ultimately we could have the College of Education. One of the interesting facts that I have discovered when working on our history is we have been in 3 different colleges. First it was Agriculture, then Arts and Sciences, then Education, and then back to Agriculture. Or, as I like to say it, matter returns to its original form. We've had 14 different deans and 13 different departmental leaders, with different titles–it wasn't always head. When it was a program, it was a program leader. We have had 88 faculty members during our entire history.

The major purpose throughout the history going back to 1918 was to produce teachers and provide in-service help to teachers, to keep them up to date. In addition, several of the majors became Extension agents, and then there's always the agricultural industry. We produced graduates that knew a lot about agriculture and knew how to work with people. That is exactly what the agricultural industry also wants. An overall purpose was working with and producing and providing help to teachers.

I would say that the most chronic challenge is the budget. When the Smith-Hughes Act was passed in 1917, a certain amount of federal money was allotted for teacher preparation. Virginia Tech accepted its share in 1918 and that's why the department got started. But, the university didn't calculate accurately. With the money they thought they had been allotted, they hired two faculty members. Then they found out they didn't get as much money as they thought they would. They had to let one go and cut the salary of the other one. We started out behind on the budget and we've been behind ever since. It's not unique to our department, but budgeting is always an issue. I sometimes say that animals in the wild wake up in the morning thinking about where they're going to find food that day and when they go to sleep at night thinking about the next day. In higher education, in public education in general, administrators wake up in the morning wondering how they're going to balance their budget. At night when they go to sleep, they think about how they're going to balance their budget, or how they're going to accomplish things the next day. It's a similar kind of thing, I would have to say that the overall biggest challenge was the budget. But in addition, as I pointed out, we also lost our degrees and those kinds of things which, ultimately, we had to get back.

Accomplishments of the program feed into the move back into the College of Agriculture and Life Sciences. This permitted us to get our own budget for the department. We had to work to do that. We had to work to get our own degrees. Which then let us get our own students. That is when we got back our departmental status. When we first got to the College of Agriculture, we had to do fundamental things before we could start thinking about making progress and adding new majors and things of that nature, because that's very basic to the department. Departments need to have their own degrees and their own students. With the help of the College of Agriculture administration, we could accomplish those things. Another accomplishment was the establishment of the scholarship program in 1980. That was the first year we had a donation. It was from what was named the Agricultural Education Society (AES)–a predecessor to the Collegiate FFA. It started out small, it was a $300 donation from AES and I said, "OK, I'll match it." Then we had $600 and got it started with that. We also started an AES/FFA leadership conference. That started in 1981 and we're now getting ready for the 37th anniversary coming up in 2 weeks. Over the years, an approximate total number of attendees was about 18,000 FFA members. Roughly 500 a year. Out on Southgate Drive, the entrance to campus, there is a gray historical marker that says founding of the Future Farmers of Virginia–getting the marker was accomplished during this time. We also created the FFA historical room.

When we first got to the College of Agriculture, we had to do fundamental things before we could start thinking about making progress and adding new majors and things of that nature, because that's very basic to the department. Departments need to have their own degrees and their own students.

This department was the largest percentage of my career and so it was very important. My first position in higher education was at Western Kentucky University and it was an across the boards kind of appointment with vocational service areas–which is what we now call Career and Technical Education. I spent 4 years there and I enjoyed it. However, agriculture and the FFA were calling me back home, so I stood back and looked at the agricultural education teacher training programs across the country and I liked Virginia Tech best. When they had an opening, I applied and got the chance to interview and was offered a position here. It has meant a great deal to me, I had a 30-year career here, so I enjoyed it.

The department has always emphasized effective teaching and good research. Some might call the research to be practical for the practitioner and we've emphasized that. We've emphasized practical preparation in the classroom, methods classes, student teaching, those kinds of things. Certainly it was philosophy woven into that, but a lot of what I took back to the classroom here preparing the teachers was based upon my visits to schools and departments across the Commonwealth. I'm just guessing, but I've probably visited 100 middle schools and high schools over the years. Every time that I made a visit, I learned something. Sometimes it was how not to do it, but generally, it was a creative idea that I could share with the student teachers and the future teachers and in in-service workshops. As one of maybe just two units in the College of Agriculture that works with social science–Ag Ed being one and Ag Econ being the other one–we worked with inter-human skills. Sometimes people would make fun of the so-called soft skills and yet that's how people lose their jobs – if they don't have those soft skills. That's something I like to point out to others. But we also worked with that. We also emphasized student organizations as one of the ways to develop those leadership skills and inter-human relations skills. I always found it interesting that when we would follow up with our recent graduates, and they were asked what were some of the most meaningful experiences you had when you were at Virginia Tech – the most common thread was the student organizations and how they got to know faculty on long trips to Kansas City for the National FFA Convention, working with the leadership conference, and things of that nature. Those are all really the social sciences, and the department certainly emphasized those.

I try to help when I'm asked in an area in which I can help out some. I am the unpaid curator of the FFA historical room and it seems like occasionally there is a question about what happened in the department and when and why did it happen. I can give some background in that. I am also a benefactor for the Hillison scholarship and endowment. Whenever the department gets stuck on a historical issue, they seem to come to me. A youngster like me, it's surprising. But, I try to help when it is needed. I also try not to get in the way as a former department head. I let Dr. Rudd handle the issues of the day but I just like to know what's going on in the department and keep up to date.

In terms of challenges, the budget issue hasn't gone away. For students in general, I see what I consider extraordinarily high tuition is a limitation. That's for all students in Virginia Tech; that isn't unique just to this department. I also have served recently on the department's promotion and tenure committee and I see increasingly higher and higher standards for our new assistant professors. I really enjoyed getting out in the world and visiting the schools and so on. Our assistant professors don't have time to do that now. They're too busy working on scholarship and grantsmanship and things of that nature. Once again, that is not unique to this department, but it is what hits our assistant professors when they're excited about being a new professor. Thinking that their Ph.D. program is the challenging part of their life then all of a sudden they get a reality. It's tough to be an assistant professor and that bothers me.

> I see increasingly higher and higher standards for our new assistant professors. I really enjoyed getting out in the world and visiting the schools and so on. Our assistant professors don't have time to do that now. They're too busy working on scholarship and grantsmanship and things of that nature.

Certainly the department and the college continue to raise funds for scholarships. The only problem is that scholarships can't keep up with the tuition. The tuition goes up faster than we can raise scholarship money. But as far as opportunity is concerned, I still see this as the major department for developing those social skills in graduates, and that will continue down the road. I see an opportunity for continuing the leadership development that is an important part for the department now, and I also see a greater expansion in the communication classes. Perhaps a minor in ag communications or something like that. Typically, the communications department at Virginia Tech has more students than they can handle. This is something that we laid a little ground work on before I retired of having a specialty in ag communications taught in this department and an internship experience. Instead of student teaching, it would be an internship at a television station or a farm magazine or a radio station. We had even worked out some of those ideas, so I would see that as an opportunity. The department has really started to get much more active in international activities, and there is certainly a lot of agriculture and education work that needs to be done in an international setting. I would see those as the major opportunities down the road.

6

Jay Poole

B.S. 1978

Professional Roles – Altria: Vice President, Communications; Virginia Tech: Director, Office of Recovery and Support; Common Sense Strategies: Owner

I am a 1978 graduate of the College of Agriculture and Life Sciences (CALS). I was an agricultural education major. Since I graduated, I have met other graduates of CALS and maintained a good relationship with many of them. I have also maintained a good relationship with many of the agriculture-based organizations, specifically including my fraternity, Alpha Gamma Rho. Some years ago, I was honored to be selected as an Outstanding Distinguished Alumnus, which connected me to other alumni and some of the people within the department. Because I am in Blacksburg frequently, I meet some of the people in CALS, like Dean Grant and Vernon Meacham, which gives me a chance to catch up with everybody.

> There is a continued emphasis on the value of FFA and the important role it plays for those in Virginia who are enrolled in vocational agriculture.

I was a student in the Department of Agricultural Education from 1974 to 1978. One of the things that has transpired throughout the years is that agricultural education specifically is no longer a part of CALS as a singular major. I could have graduated in the College of Education or CALS, and I obviously chose to graduate from CALS. But at that time, there was a College of Education, and I think that some of the agricultural education majors did graduate from the College of Education. One of the things that has remained constant over the years is a specific emphasis on the agriculture part of the major, which is important. There is also a continued emphasis on the value of FFA and the important role it plays for those in Virginia who are enrolled in vocational agriculture. This is a significant part of the legacy going way back before I went to school and what agricultural education at Virginia Tech is.

At the time I was a student, the agricultural education majors were focused on training to be instructors. There never seemed to be enough agriculture teachers in the Commonwealth of Virginia to meet the demand of the school systems. Certainly, the primary effort of the Agricultural Education Department in those days was to provide a steady stream of teachers who go into the secondary school system of the Commonwealth. Some of us who chose other career paths besides teaching and, in fact, there were probably as many who chose a different career path as those who chose to become ag teachers.

As I recall, there was a significant challenge, because the program was primarily, for administrative purposes, located in the College of Education. That was a disconnect, because most of us were more agriculturally oriented and most of our classes were in CALS to meet the degree requirement. That disconnect between having this education degree with a very specific agriculture component sometimes made course selection a little more complicated than it had to be. With that said, a lot of credit goes to my advisor – Dr. John Hillison – with whom I still maintain contact. John made sure I got the right classes and took the things I needed to take to stay on course to graduate. In my case, I was very lucky that I had a good advisor who helped smooth out those edges, and I think that made what could have been a challenging experience a lot easier.

I recall the emphasis placed on the connection between vocational agriculture and FFA. Because the FFA started in Virginia, I think the importance attached to the FFA legacy, as a piece of the agricultural education department, was significant and important.

One of the things I enjoyed was that, as an agricultural education major, you could take a variety of classes in CALS. To me, that was one of the best parts of it, because you did not get stuck in an animal science curriculum where you take a lot of science that you may never use or get stuck in an agricultural economics curriculum with a lot of economics theories that you may never use. You got a chance to take a broad look at the curriculum areas. That was attractive to me, and I think I had a broader academic background as a graduate because of it.

I had a really good experience as a student teacher, both with the students and the supervising teacher. The experience was so good that I knew it had to come down from there, so that's when I decided I didn't want to teach anymore. There was no way a real teaching job could be as good as my student teaching experience was! The curriculum that went along with an agricultural education major provided a breadth of different kinds of academic experiences within CALS to provide a real taste for a lot of different things. And certainly, if you consider that the primary objective of the agricultural education department was to give graduates what they needed to become agriculture teachers – if you think about it in that way – then the broad exposure to many aspects of ag in college provided a very good platform upon which future teachers were exposed to different aspects of agriculture and which they could then use in their own instruction to high school students.

I am on the departmental mailing list and on the email list. My sister, Billie Jean Elmer, is also an agricultural education graduate, and so, through her, we keep up with what is going on in the department, and if I choose to engage, I can. In that sense, the ongoing communication that the department has with its graduates is good and at some point I might choose to engage.

How do we explain the value of the degree to potential students...where fewer people aspire to be agricultural education majors.

I think as Virginia becomes more urban and suburban, encouraging CALS student to become agricultural education majors poses a challenge. Especially since the value of an agricultural education degree, perhaps beyond being a secondary school ag instructor, doesn't immediately jump out and grab you. I think that may be one of the challenges, how to explain the value of an Ag Ed degree to potential students who want an ag degree, where fewer people aspire to be agricultural education majors.

I want to add that we are talking now 40 years later. When I was in school, a lot of the agricultural education majors were farm kids themselves. And so, you had students with a more natural affinity to gravitate towards agricultural education, particularly if they were engaged with FFA in high school. You were more likely to be predisposed to think of agricultural education as a major. Whereas now, there are fewer farms and so fewer high school students in rural Virginia, a natural affinity which existed for agriculture and exposure to agricultural education as a potential major is significantly less than it was 40 years ago.

I see opportunity for the department in the need for education beyond traditional "Agriculture Boundaries." When I was working at Altria, we started a program called Ag Relations. At that time, we were a large consumer products company. The thing that is unique about every single one of Altria's consumer products is that all of them were agriculture-based. Even though the company was a large consumer of all kinds of ag commodities, there was no visible link between the farm and consumer. We created a department, conducted some research and found out - not surprisingly – that farmers are the first to say, "Nobody likes us, nobody loves us." When in fact, that's just not true. In our research, we found that only firemen were more respected than farmers. What we also found was that consumers did not understand farm practices. It was not that consumers did not like or respect farmers, but they did not understand anything about the products they produced and how they produced them.

So, I think the opportunity that exists for agriculture and agricultural education is to help consumers better understand the food system and how food and fiber comes from the farms and forests and finds its way to the consumer. Agriculture teachers would be uniquely qualified to create and use an external educational platform to address erroneous misconceptions of agriculture which widely exist among consumers. Some of these ag myths are preposterous, but once a misconception takes hold, it is difficult to stop it. I think that is a wonderful opportunity for agricultural education to take a lead role in helping to identify ways that agriculture can be more effective at telling its own story. Instead of just preaching to itself.

7

General Carlton Everhart

B.S. 1983

Professional Roles - Air Mobility Command: Scott Air Force Base, Commander

I graduated from the department with a degree in agricultural education and an emphasis in agricultural business in 1983. I was looking into a degree in agricultural business. It was a brand-new curriculum that was placed in the College of Education. My plan was to eventually go into farming. I closely associated with the hard work and discipline of the farming community and wanted to help out my family.

At the time, it was a brand-new curriculum and the university was trying to balance out where the agriculture business piece and the agriculture education piece of it would fit. That finally got settled, or close to being settled, in my senior year. Overall, the curriculum was solid and tight - it offered a great foundational academic baseline with broad applicability to not only agriculture, but to business as well. It definitely enhanced my problem-solving skills and ability to teach and mentor others.

(T)he biggest benefit I gained from the department was learning how to best engage people and how to get the most out of each individual.

The education component was the primary purpose of the agriculture department at that time. For a new program, challenges included figuring out the new curriculum, where it would be placed and what colleges it would be placed under. As a student, we didn't get exposure to the inner workings of the department for the most part, but the curriculum was challenging and provided cross-discipline exposure. I felt the department was leading the way for the type of education provided to students, while also offering insight into the business elements.

As for being an Air Force General Officer and what I took away from the Virginia Tech experience, I felt the biggest benefit I gained from the department was learning how to best engage people and how to get the most out of each individual. In my career, there were times I had to instruct pilots, so the education piece and how to get the most out of a student can be directly attributed to what I learned at Virginia Tech and the department. That was a huge benefit for me, because there were many aspects of the curriculum that gave me the skills that I needed to maximize developing and mentoring students and Airmen during their Air Force careers. In a way, the education not only benefited me, but also enhanced my ability to help develop the skills of others.

I feel the primary way my work in the department has influenced the lives or careers of others is that it set a baseline for me as an instructor pilot. I was able to fall back on the experiences and practice that I gained during my time at Virginia Tech. Now, I hope this doesn't come across as bragging, but I believe that my students did well based on the teaching guidelines and techniques that I was provided and exposed to in college. Virginia Tech helped me discover ways to effectively teach and get maximum productivity out of every student. Virginia Tech taught me, and in turn, I was able to teach others and pass along the valuable lessons learned. When it came time for my students to pass their exams, they did well. When there were problem students, they were always handed over to me. Is that a blessing or a curse? Well, I think it's a blessing, because I felt I could always help a student. My goal was to always get a student over the finish line and help them achieve their full potential.

Relationship-building and continually looking to improve oneself is critical in any line of work.

My times at Virginia Tech are filled with positive memories. I'm not able to be involved in the department as much as I would like due to commitments to my airmen and Air Force career. But I continue to value my time at Virginia Tech and want to give back – either from a leadership or teaching perspective – whenever I can. I feel it is important to pay forward what was taught to me both at Virginia Tech and in the Air Force. I have a lot of respect for Virginia Tech and the agriculture community. Both are committed to hard work and provide a very valuable service to our nation.

Unfortunately, I think right now, as corporations get bigger, the small farmer is facing greater challenge to make a living. I think that's a challenge for the department today overall, because that uniqueness is what sets the conditions for people wanting to go into the business. As we become a more service-based economy, and as we export more; the emphasis on agriculture, I personally believe, is losing some of its glamour and roots. I think that needs to come back as an area of focus for America. Farmers are some of the hardest working people in America. That is a trait our Airmen and farmers have in common.

The department has an opportunity to provide students tangible experiences - whether it's Future Farmers of America or "learning by doing." It is important students are exposed to successful models and gain valuable hands-on skills. That's where the greatest benefit in agricultural education and agriculture itself is – learning by doing, establishing your relationships, and setting the foundation for productivity. Relationship-building and continually looking to improve oneself is critical in any line of work. It is the key to success in the environment that we're in today, as well as in the future.

I am grateful for the experiences I gained at Virginia Tech. They afforded me exposure to different ways of problem solving, the value of developing strong relationships, and how hard work, innovation, and dedication are fundamental to success in agriculture or in the Air Force. I recognize this and will forever be grateful for the experience.

8

Barbara Kirby

Ed.D. 1984

Professional Roles - Ashland County: West Holmes Joint Vocational School; Ashland, OH: Agricultural Education/Environmental Science Teacher; Virginia Polytechnic Institute and State University: Graduate Teaching Assistant/Instructor; The Ohio State University: Lecturer; North Carolina State University: Faculty, Assistant Director of Academic Programs, Associate Director of Academic Programs, Director of the Agricultural Institute, Associate Vice Provost for Academic Programs and Services

I was affiliated with the Agricultural Education program as a graduate student from 1982 until 1984. My position was a half time instructor. In 1984, I graduated with a Doctor of Education degree. The University was called Virginia Polytechnic Institute and State University, or VPI. Now, my affiliation is to be a good alumnae, follow the Hokie sports teams and network with the faculty. I attend events in Blacksburg or other places when I am available.

I earned a Doctor of Education degree. We were administratively housed in the College of Education, as it was known at that time. The Agricultural Education program was part of the Division of Vocational and Technical Education. My major program area was agricultural education (teacher education). My cognate was rural sociology. The program is different today in many ways, including its administrative home.

The agricultural education program was in the Division of Vocational and Technical Education (VTE). It was a vocation education division that included other Career and Technical Education (CTE) programs. Not a department of Agricultural Education.

In agricultural education, most of the program focused on teaching and learning in schools or non-formal education in extension. There was not a focus on leadership or community engagement. Extension did have a role. If some of our students decided not to teach, they might work in agribusiness or perhaps work in Ag. Extension. Eventually, the student needed a masters degree.

When I was in the department, it was a fairly stable department, in terms of an Agriculture and Extension Education program. The faculty members I worked with when I came in 1982 were still on the staff in 1984. One of the biggest changes in personnel was that Dr. Crunkilton had a sabbatical in Greece working with the Farm School. He was gone about six months while I was at Virginia Tech, but I still had an opportunity to work with him. We also benefited from his educational leave. The people he was working with at the Farm School in Greece came to Tech. We were able to meet some of those individuals and had the opportunity to know more about what they were doing at the Farm School. The program ran smoothly while he was away. Dr. Crunkilton was always in communication with us. I believe Dr. Clouse stepped up to assume some of the day-to-day leadership duties. Faculty members take educational leaves so others step in to help the program run smoothly.

People were starting to think about the administrative home of ag ed programs. Should the program be located in education or agriculture? Where are resources?

I believe there was also some discussion at the time about the structure of the overall department in VTE, the School of Education, College of Human Resources, and how that would all evolve. However, that was typical of higher education around the country. People were starting to think about the administrative home of ag ed programs. Should the program be located in education or agriculture? Where are resources? What are the faculty strengths and collaborations in education or agriculture or other administrative homes?

I think the purpose of the program was primarily to prepare students for careers in Agricultural and Extension Education, such as teaching, working with Extension and agribusiness. Students may have also gone to some farm, commodity group, or organization.

My work with the program focused on undergraduates. I helped with the Ag Ed Society and the honorary fraternity, Alpha Tau Alpha. I taught introductory level courses and supervised student teachers. It was a good-sized program. I can recall that in terms of national visibility and ranking, the program was well respected and always has been viewed as a prestigious program by our peers.

Ag Ed was not a department, but part of a division of VTE. The structure provided opportunities and challenges. I think it's always a challenge when your program resides within a more diverse department or division. We had a division that included marketing education, business education, and other vocational programs. The structure provided a unique way to collaborate with people from different program areas. You had the opportunity to do more of vocational education type of work and research. There was a lot of opportunity to be a national leading program in vocational education as a result of being part of education and collaborating with national vocational education centers. I think the program had good opportunities. There were great, nationally known faculty members across the division. I had friends who were in other programs in the division at the time.

The program continued to have a strong undergraduate and graduate program while I was there. I worked a lot with people on their specialty areas. The program had a strong historical focus. Faculty members, including Dr. Hillison, Dr. Clouse and Dr. Camp, all valued our historical foundations. We had many of the artifacts and early books, as well as the table where the founders of FFV and FFA gathered. My fellow graduate student, Mike Rush, and I had plenty of opportunity to review historical materials along with hauling the "Table" to National FFA conventions. I was interested in the pictures and materials we found from the time when Henry Groseclose and others started the FFA. There was always work to be done in the historical archives. We were trying to get more and more of those items cataloged. I think that was a long-term project that Dr. Hillison still works on today. I teach a History and Philosophy of Agricultural and Extension Education course today that is similar to the courses I took from Dr. Hillison and Dr. Camp. Dr. Clouse and I worked closely with the National FFA Building Our American Community program. I was fortunate that my dissertation work focused on FFA Chapter community development projects and was supported by FFA. The Tech program was very supportive of state and national FFA. We were very involved with the teachers and the FFA chapters and organizations.

I can remember bringing the first female FFA president, Ms. Jan Eberly, to Virginia Tech in 1982.

One of the other accomplishments was integrating technology, particularly instructional technology, into Education and Agricultural Extension. Dr. Bill Camp was my adviser, and we worked with the teachers. We developed computer literacy workshops. At the time, it was the beginnings of the Apple IIE, the Tandy's, and the Radio Shacks and some other relic desktop computers. I collected data for my dissertation and punched cards one semester, ran tape and used a terminal the next and finally completed my dissertation on about three 5 ¼" floppy discs. There was a lot of technology change in two years. Changing hardware and software impacted the entire division. The schools weren't using the computers. People hadn't really been trained. The opportunity was available for us to develop lesson plans and workshops to help teachers understand how to use the computers and how to integrate instructional technology into their classrooms.

Teacher-education was my primary focus, so I supervised student teachers out in the schools and worked with their cooperating teachers. I traveled out from Blacksburg a couple times a week to visit the student teachers and work with cooperating teachers. The Ag Ed program graduated competent, talented teachers.

The relationship between agricultural teachers and the department was very good. We participated in their various annual conferences. We were very active with the VA FFA Convention. I can remember bringing the first female FFA president, Ms. Jan Eberly, to Virginia Tech in 1982. She spoke to the Virginia FFA Convention That was quite an exciting time. Her speech was memorable. Jan was delivering a great speech, but appeared to get blinded by the stage lights. She walked right off the stage into the VA FFA tympani player's arms. He set her back onto the stage. She didn't miss a beat and just continued with her talk. Needless to say, she received a standing ovation. I also really enjoyed being part of Omicron Tau Theta, a national professional honorary organization in vocational education. We had a very strong local chapter. I served as a local officer and with other faculty on the national level several years after I graduated.

My role was to be somewhat of a teacher educator in training. I enjoyed teaching some classes. I enjoyed supervising the student teachers and their cooperating teachers. I enjoyed, when I had an opportunity, to work with the Alpha Tau Alpha, an agriculture education honorary for undergraduates. I was preparing to become a professor in a faculty position for teacher education. I continued to work closely with the National FFA and served on the Teacher Educator Board of Consultants and as a member of the FFA Board of Directors. Living in VA and learning about the FFV and history of FFA definitely made an impression on me.

And now, after spending about 17 years as an administrator, I'm back as a teacher educator/faculty member. I think everybody takes some time to grow and to do some different things. And that's why I went into administration. Administration wasn't in my program courses, other than a couple of courses. I had enough courses in order to get a principal's license and a vocational director's license, but not necessarily preparation for university higher education administration. However, I had good administrative role models at Tech.

Now, in returning to the faculty, I'm doing the same work that I was being prepared to do from 1982 to 1984. I've come full circle.

I hope that for those students who perhaps had an interest in using the computers that they learned something about instructional technology, without fear and anxiety. They were able to integrate computers into their instruction. Teachers, in particular, were just getting comfortable with starting to use the computers in their classrooms. I hope I helped. When we supervised the student teachers, we always looked at the type of technology teachers were using – overhead projector or computer? Could they do some other kinds of things to make their work more efficient and effective? Hopefully, the instructional technology would help them save some time unless they weren't very good typists and they were typing with both index fingers. I think in terms of careers, we really tried to encourage and support students. If they were having trouble in a class or challenge with a project, you'd be there to try to help them understand what they were supposed to do. When you saw them go across the stage and they graduated, then you were successful. Graduation rates are still important today as a measure of student success and launching their careers.

My current involvement with the program is limited. If the department has some type of a function, I generally will get involved with that. The department or college always has a tailgate and hosts one of the football games in the fall. I try to reconnect, because it's really about maintaining those relationships and finding out what's happening in the department. I want to know how's it growing, how's it changing? When we moved from Lane Hall to Litton-Reaves, that was a tremendous change for the department. More faculty had opportunities to be part of this department in a College of Agriculture. We can now attract a wider audience of students with many more interests than what we had when we were a more narrowly-focused program. The program has really broadened its mission. Even the name has changed: The Department of Agricultural, Leadership, and Community Education

Dr. Rudd and I were just walking around up here and just talking about the number of faculty offices, and how many faculty are here now, and all the degree and program change. I spent some time reading about the department on their web page. I was impressed by the different types of roles that faculty have now. The different expertise that they bring to this department is impressive. So it's a much more diverse department in many ways. The department is more diverse in terms of gender, ethnicity and disciplines. That wasn't the case when we were back in 1982.

As of the 1980s in the Ag. Ed program, there were only one or two other female graduate students enrolled. We did not have women faculty in Ag. Ed, but there were several women faculty members in the Division. However, I fondly remember Doris, our secretary. She was very supportive of me while I was at Tech. There weren't very many females in agricultural teacher education, and so you didn't see many females on the faculty around the country. In 1980, '81, '82, women were first being hired to be teacher educators, higher ed faculty members in ag ed departments. There was a lot of tradition and a lot of difficulty in breaking through some of the traditional expectations of what is an ag program and who should be employed at the university ag ed program. Agriculture education was mostly boys in the high school; we called them boys, and finally, girls were admitted. And then some of those female students went to enrolled in ag ed programs. During my own undergraduate program and as a teacher, I was the minority gender. Women represented about 3-5%. Now, it's predominantly women enrolled in undergraduate agricultural education. We are more diverse. The diversity is not really where we would like to have it. I also commented to Dr. Rudd about how wonderful it is to see the diversity that you have here now among the graduate students. I think that is really special. That is a special impact that the department has had. It's a big growth area that the department has experienced. So I really congratulate the department on their efforts in growing and diversifying the graduate program.

So I really congratulate the department on their efforts in growing and diversifying the graduate program.

I think one of the biggest differences now is that the department is able to offer the Ph.D. As a part of the school of education, our degrees were in education. And I think that a number of students wanted to earn a Ph.D. You could get a Ph.D. at the time I was in school, but you had to take an additional statistics/research courses. We still had research and statistics courses in our Ed.D. Programs; the students wrote and defended a dissertation just like the Ph.D. Most of the Ed.D. students were practitioners in education or extension. I think that the department has had to do quite a bit of curriculum revisions in the transition. When the program moved to the College of Agriculture, the department was able to offer the Ph.D. degree.

Perhaps there could be some student exchanges. Maybe a student from Virginia Tech would like to spend a semester at NC State and an NC State student go to Tech. We could facilitate a faculty member, or a faculty exchange, or a student exchange. I think that would be kind of a neat idea. We don't do the national exchange as much as we used to, where students could go to another university within the United States and you still pay your tuition at your home institution, but attended a different institution. A student might be interested in learning about a particular program at NC State, or we might have a student who's really interested in learning more about the leadership or the engagement programs that Virginia Tech offers. We could also explore doing more inter-institutional collaboration with co-PIs on grants. That would be another way for alumni of the department to be more involved with the department. As partners, we would bring our areas of expertise to grants addressing common issues or challenges. I think there are always ways that a faculty member, or a graduate, can become involved again with the faculty at their home institution. I also try to contribute to the scholarship programs.

I think some of the greatest challenges may have to do with the budgets that we have available to higher education. We continue to face challenges in delivering our programs that are impacted by the budgets in higher education. Another major challenge is helping agricultural students access our higher education institutions. Students are great today. They have really high ACT and SAT scores and high school GPAs. It's very competitive. We worry at my institution, as I know Virginia Tech does, as well, if students from rural areas still have access to a college education at Virginia Tech? What are the pathways that students might have to follow in order to be admitted to the institution? The faculty are spending a lot of time writing grants, getting money to support program and graduate students. Those kinds of responsibilities for the faculty may mean that they have to buy out their time and hire others to teach some of the courses.

The department has broadened its mission in agriculture and extension education, leadership, and community engagement more than the 1980s faculty even dreamed about.

I think the department has a tremendous opportunity because it has broadened its mission in order to serve more people. ALCE can make bigger impacts in so many more areas. We did not even think about some of these areas when I was here in 1982. When you broaden the mission to include leadership and community engagement, along with redefining your programs as agriscience and STEM, your work is bigger than teacher and Extension agent preparation.

There's going to be a super 100 year anniversary celebration for the department. As the department looks back over the past 100 years, examines the kinds of things that our founders did 100 years ago and how we developed and expanded upon those initiatives, there will be a multitude of reasons for celebration. The department has broadened its mission in agriculture and extension education, leadership, and community engagement more than the 1980s faculty even dreamed about. I think that the opportunities will evolve from the creativity and the drive of the faculty and the graduate students who are part of this program. What new areas does ALCE want to research? How will their programs impact others, in Virginia, across the country and internationally? What kinds of audiences do they want to serve as we look at the next 100 years? I think that's where the opportunities are going to be. It's difficult to predict what's next. For example, we can now visit student teachers remotely through technology. Every time there's a new technology, it seems like we have a new opportunity to do something a little bit differently.

For example, I was watching a high school teacher teach her unit on technology and agriscience and she was talking about the drones, precision farming and emerging technologies in agriculture. We need to incorporate instruction about emerging technologies into the lessons for students who are going to go out and teach these kids who may be staying on their home farms or other agribusinesses. The department has an opportunity to make a difference in the climate of the campus. I envision more faculties engaging with partners around the world to address our greatest challenges.

Finally, when I was here in '82, I was a consumer. I was the student and the faculty members were my mentors. The faculty all brought different kinds of expertise, and advice, and coursework, and experiences to my program and helped me develop as a teacher educator and as a person. As I moved through my program, we became colleagues. I value those collegial relationship and know those relationships are still there today. I have wonderful networks. Many of the current faculty members are people that I work with at professional conferences or in other capacities. We may share a research interest, or as I said earlier today, a source of advice.

The last thing I would say again is thank you. I'm looking forward to the 100-year celebration. I think there is so much rich history here at Virginia Tech and in this department, including the FFA and the Corps of Cadets and all the many traditions that are associated with the institution. We have a rich history, including people who have been in agriculture education throughout the years and who have really been pioneers in the field. Looking back, it will be fun to examine the past 100 years and dream about the next 100 years. I can't wait to celebrate on September 22nd.

9

Mike Rush

Ed.D., 1984

Professional Roles - University of Idaho: Agricultural Science Teacher, Instructor; Penn State University: Assistant Professor, Idaho: Ag Ed Supervisor, Director of Research and Head of Career Technical Education; Executive Director of the Board of Education; South Dakota: CEO of the Board of Regents

I was a half-time instructor, and a doctoral graduate student in the department. I started that in July of 1982 and finished in July of 1984. The Ag Ed Department was in the Division of Vocational and Technical Education in the College of Education. The primary goal was developing new teachers for high school agriculture programs and also providing professional development opportunities, including advanced degrees, to teachers and others who needed those credentials.

The department and the faculty always seemed to be early adopters of technology. I know Martin McMillion had one of the first Apple II computers and had had one for some time. He was using it to develop Ag business curriculum for active farming operations, which, at that time was very innovative. Some people kidded him about the limitations of the technology (particularly compared to the university mainframe) and suggested that the technology would be better if he just waited. He would always reply, "Yeah, but then I would have not been using the computer for two years either." We also used BITNET, which was an early precursor to the internet, and used to share research data among universities.

It seems like some of the challenges at that time are the same challenges we face today. Part of it was striking the right balance between the content providers, i.e., the College of Agriculture, and the pedagogical role played by the Division of Vocational Education. The College of Education itself was not always the best fit for the Ag Ed program area, because its focus was more elementary and standard academic kind of preparation. Having the various technical programs together, however, was of great benefit to me as my career developed. At the broader university level, that was a time when budgets were tight and universities were looking for ways to cut costs, including reducing faculty. In fact, one of the studies I did as part of my MBA minor focused on tenure and whether that was the best model for organizing the university workforce.

It was kind of an interesting study, because I had gotten permission from the university administration, the president's office, to conduct the study, because I needed access to all the faculty email addresses. The questionnaire focused on tenure and alternatives to tenure, such as fixed-term contracts. Right before I was going to send the survey out, I got a call from the university administration office saying, "You're not authorized to do this study." I said, "Well, shoot, it's kind of late in the process to make that decision," and they said, "Oh, no. We still want you to do the study, you're just not authorized. If you felt like sharing the results with us that would be fine." They were just so fearful of getting any kind of message out that the university administration was looking at tenure. The survey showed that a majority of the tenure-track faculty and a significant number of tenured faculty actually preferred a contractual structure other than tenure.

> I learned not to be afraid of new challenges, even if they seemed at the outset pretty insurmountable.

The faculty and the department at that time, and I think still do, had an excellent reputation for graduate programs. The department was a leader in developing the agricultural business curriculum. The program head, John Crunkilton, had just completed a new book during that period focused on the problem-solving method of providing instruction. The other thing I thought was impressive about the department is that it had a very egalitarian culture based on merit and effort. I remember one time the students in the college of business were conducting a study in organizational relationships and the power in those relationships. And as part of the study, they were measuring the distance – they actually had measuring tapes – and they measured the distance from the faculty offices to the department head office as a measure of power and influence within the department. I told them, "I am afraid your data collected from this department is going to screw up your study." My office and those of my fellow instructors was right across the hall from the department head. We had a big, large window overlooking the tennis courts. I'm sure our profile did not fit their expected model.

Another example was that the department chartered a bus to go to the professional conference, rather than paying for just a few of the senior faculty to make the trip. That allowed all the graduate students and everybody else to go along as well. That attitude just provided excellent opportunities for younger faculty to develop professionally.

My role in the department was certainly key in influencing my career. It provided the necessary skills, credentials and experiences to get my first full-time faculty position at Penn State University and then increasingly more complex roles as head of career technical education in the state of Idaho and, ultimately, the executive director of the board of education for Idaho and also the executive director and CEO of the board of regents for South Dakota.

I learned not to be afraid of new challenges, even if they seemed at the outset pretty insurmountable. I remember Jim Clouse calling me down to his office and saying, "Mike, I really need you to take this new responsibility on. They just got a new grant and they need somebody to kind of head up the effort." I didn't think I had the expertise that they needed, but he just said, "You know, sometimes you just have to buy the pig in a poke and make it work. You just have to jump in and see how it turns out." That turned out to be extremely good advice, because I faced a lot of those circumstances later in life.

Another thing that I learned was to take advantage of the opportunities available, rather than waiting for the perfect time to do something. I remember one time John Hillison and I were working on a study. We were trying to get it done in a fairly quick time frame. We worked out all the questions on the bus ride to one of the professional development conferences, and then we validated the questions by passing the cards down the line to all the faculty on the bus. I just thought, "Well, that is not the way I imagined organizing a study, but it certainly took advantage of resources that otherwise would have taken weeks to marshal."

I also learned that it was necessary to connect solid theory with solid practice. One of the real watershed moments in my teaching career with regards to that particular topic was when I was assigned to teach a class in how to teach adults. It was an evening class targeted to teachers. During the very first class I'm looking out, and it dawns on me that all the people in the class were adults. I'm teaching a class of adults on how to teach adults. If what I say doesn't jibe with what they experience, this class is not going to go well. It was extremely instructive to me that it wasn't only important to know the theory, but to also make sure you could translate that into actual impact and practice.

And I also learned that regular people can have a huge impact on major issues. As part of my doctoral program, I got to do an FFA Alumni internship in Washington, DC. It consisted of visiting senators and Senate staff and White House staff, lobbying on behalf of new Perkins legislation. I was amazed to see how well people listened to even a graduate student from Virginia Tech.

My career has certainly benefited from my experience at Virginia Tech. I have had the opportunity to play a role working alongside hundreds of talented and dedicated people in providing educational opportunities and success to thousands of students at both the public school and college levels.

I think challenges for Agricultural Education and Extension are maintaining relevance in the changing economic, technological, and cultural environment that we face. Agriculture is, in a lot of ways, becoming increasingly disconnected from the grassroots farming operations. Many of the jobs associated with agriculture are done far from the fields and feed lots – economic modeling, DNA sequencing, marketing and exports. Fewer and fewer kids grow up on a farm, and they don't have the same sorts of opportunities and experiences that defined agricultural education in the past. I just think that agricultural education programs will have to continue to innovate and modify instruction and programming to bridge those gaps.

I think one more challenge is maintaining strong teacher and extension ed preparation programs in the face of reduced resources and the increasing cost to students. Cost for higher ed is getting out of line with what students can afford. That is going to put increasing pressure on higher education to figure out new ways to control costs and still provide quality education. If we don't, people will look for other strategies for their career development.

I think, frankly, Agriculture and Extension Education has the robust curriculum that appeals to students from a variety of backgrounds with a variety of interest. It's a hands-on curriculum, but it's also very academic. And the leadership development is particularly inviting to a broad range of students, no matter what their background. I think that robust curriculum, as long as we continue to nurture and adapt and take advantage of it, has real long-term potential for keeping our programs relevant and valuable to our high schools.

I deeply appreciated my time at Virginia Tech. I don't regret any of the experiences that I had. I had exceptionally great faculty to work with. And, as I said before, the culture of the program was not to relegate graduate students or even the junior faculty to some back room sorting pencils. The culture encouraged meaningful involvement in the entire operation of the department. That created a tremendous educational opportunity.

10

Regina Smick-Attisano

Ed.D. - 1988

Professional Roles - The University of Maryland: Assistant Instructor, Institute of Applied Agriculture; Assistant to the Director, Institute of Applied Agriculture; Assistant to the Dean of the College of Agriculture; Maryland Governor's Commission on Education in Agriculture: Commission Intern; Virginia Tech: College of Education – Assistant to the Faculty President; Agricultural Technology Program - Instructor; Academic Advisor and Instructor; Coordinating Academic Advisor/Assistant Director and Instructor; College of Agriculture and Life Sciences – Acting Assistant Director of Resident Instruction and Instructor; University of New Hampshire: Director and Associate Professor of Thompson School of Applied Science; Associate Professor of Thompson School of Applied Science

I was a graduate student pursuing my doctoral degree - I believe the second female one, beginning in September 1986. John Crunkilton was my major advisor. Virginia Tech was still on the quarter system then. I finished in December 1988. John Crunkilton was the College of Education Faculty President at that time, and I was his assistant. I'm not sure exactly what the name of the department was at that time. However, I know my degree was in Vocational and Technical Education. We were a part of the College of Education.

> I do know it was an amazing graduate student experience. I made the right choice of attending Virginia Tech.

I don't recall any significant changes occurring during the time I was affiliated with the department. The primary purpose of the department at that time was to prepare undergraduate students to become agricultural education teachers at the secondary level, as well as masters students, along with preparing doctoral students to become teacher educators. We supported Virginia agricultural education teachers and the Virginia FFA Association as well.

I don't remember any challenges for the department, but I know at some point the department was trying to get the Ph.D., as well as the Ed.D. degree, available for students to pursue. I believe the program grew in numbers during the time I was there. My assistantship was an administrative one, which was quite unique. I do know it was an amazing graduate student experience. I made the right choice of attending Virginia Tech.

I hope I was a role model for more women to enter the profession of agricultural education.

In my career, I held the majority of my positions with either an administrative appointment that was halftime and/or full time. My longest administrative post was for 20 years at a postsecondary institution, preparing students mostly for agricultural careers. I hope I was a role model for more women to enter the profession of agricultural education. Some of my role models for being an effective administrator came from my association with Dr. John Crunkilton, Dr. Jim Clouse, and Dr. Frantz at the time I was there. I have always worked for associate degree programs offering agricultural disciplines and where excellent teaching was the standard, and I believe my doctoral education from Tech was a strong underpinning for my success in strengthening the teaching in each place I worked.

I am currently a strong supporter, as an alumna of the department. I receive the CALS newsletter and try and stay in touch with Rick Rudd, who was a graduate student after I finished.

The biggest challenge facing Agricultural Education and Extension is society's view that agricultural careers are not viable. The greatest opportunity for the program is to continue to grow and advance the causes of the need for college educated individuals in the variety of agricultural disciplines and careers.

11

Rick Rudd

Ph.D. 1994; Faculty - 2006 - present

Professional Roles - Virginia Tech: Interim Associate Dean and Director of Virginia Cooperative Extension; Agricultural, Leadership, and Community Education Department Head & Community Viability Chair of Excellence Professor

I'm currently the department head, started in 2006. Looking to step down from this role and go back to faculty, hopefully in January. I actually did my Ph.D. here in '94 and went to Florida for 12 years and came back. I was hired as department head.

When I was a graduate student, we were a program in Career and Technical Education (CTE). And so the department really only came about in the early '90s, I believe, before I got here, and it was Agricultural and Extension Education. And so we've changed it since to our current name, with a lot of work and a lot of input from stakeholders and everybody else on the university campus. It's a big deal to change the name of a department. Doesn't seem like it should be. But, the current name was one that the faculty really identified with, with all of our programs. CTE is now in the school of education, when I did my Ph.D. here, it was one of the biggest and best programs in the country. And so actually my Ph.D. is in CTE. Vocational-technical education is what it was at the time.

Right before I got here as a student, the department left CTE, the College of Education, to come to the College of Agriculture. The College of Education was not happy, but the Ag Ed program had really been stifled there, and this was a positive move to get Agricultural Education into the College of Agriculture and Life Sciences. Then, in the early '90s, the Extension program development and evaluation group was added to the department in Ag and Extension Education. We've had Extension program evaluation people since then in the department, which is really kind of a legacy of that move.

When I got here in 2006 we had two Ph.D. faculty, one tenure-track faculty member and me as a tenured professor and department head. The goal was to grow the department and grow our areas of emphasis. The department was primarily a teacher education unit that had gone down to just one faculty member in teacher education. We were preparing teachers and that was it. I came from the University of Florida, where we had a much more diverse department. And at the time, you could look around the country and see departments closing left and right. And in the modern university, you can't have a department that's just a singular focus. When I interviewed here, I told the dean at the time, this is not going to be sustainable as a department. You could be sustainable if you diversified – you added community development, you added more in the Extension area, you added leadership, you did some other things along with Ag Education. She bought it, so I came, and we've been working on that ever since. So we have 18 faculty now, tenure and non-tenure-track folks. We didn't have a grad program when I came. We've grown one of the best grad programs in the country. The bachelor's degree in Agricultural Sciences was approved by SCHEV in the spring of 2006. Our undergrad program is still small. But there have been a lot of changes over time. We've hired some great folks in Extension, and in leadership, and in community development. We've really grown those areas well.

The goal was to grow the department and grow our areas of emphasis.

The primary purpose of the department was training teachers, and we still do that. And I argue that we would not be doing that today had we tried to stick to that singular focus. Programs like Maryland are gone. Cornell is gone. Penn State's Ag Ed unit has been rolled into another department. North Carolina State's Ag Ed department has been rolled into another department. So you look up and down the East Coast, and it is Virginia Tech, University of Georgia, and University of Florida, and those are the only three departments like ours. And those are all three diverse units that have lots going on. So I think that was a smart move for us, and we really sell ourselves as a social science unit in the College of Ag and Life Sciences. We're the people department, and that's important. And I think we've become a very important component of the college's broader mission because of what we add to other things going on in the college with the human side of agriculture. Our department is the human component, the human element in agriculture and life sciences. We know things about education. We know things about program development and evaluation. We know things about leadership development. We know things about human and institution capacity building. We know things about viable communities. I mean, those are all pieces of who we are now as well as agricultural education.

We still have some constituents who are not happy that we've changed who we are. But I think over time, that will – it has smoothed out a lot and that will continue to get better. We've got to do a better job of communicating who we are and what we do, because I think that's probably the biggest – our biggest hurdle is we don't talk enough about how we are helping. We have a doctoral defense today that's a case study of a program we've been working with in Carroll County for a decade in helping them build a modern ag education program. There was just a news article about it. It's been featured on CNN. Former governor, Tim Kaine, presented it to the Senate about two years ago. This is going on in Carroll County, Virginia. So we are doing some really cool work and being a catalyst for change, and I don't think we talk about that enough. That's what a university should do. It's not perpetuating what's always been there. It's always kind of, what's the cutting edge? How are we going to get better?

We've got to do a better job of communicating who we are and what we do because I think that's probably the biggest – our biggest hurdle is we don't talk enough about how we are helping.

A couple of things that I'm really proud of. We've really made a push to be a diverse program at the graduate level. We purposefully recruit and prepare underrepresented students in our field. We probably do that better than anyone in the country or anyone in the world, for that matter, and I'm really proud of that. That's been a push that we've had. The second thing I'm really proud of is we truly have a world-class faculty. We have faculty here that are known internationally. A lot of relatively young people, but they're working really hard and making a big impact. And so I'm proud of our faculty, and I'm proud of our students.

I've always been happy in my career. I was happy as an agriculture teacher. I told people that the hardest professional decision was to leave that and go back to graduate school. It was a very hard decision to leave Florida to come here. I'm really happy here. Stepping down as department head has been my plan all along. I thought I would do this department head thing for about 7 years – it's been 12 – but then go back to faculty. I'm really looking forward to being a full professor in the department, Teaching, doing my research, doing my extension work, and contributing as a faculty member. So hopefully, I can be one of those great faculty that we have and do some cool things to make a difference.

I love mentoring the faculty, especially young faculty going through the assistant to associate. So every time one of them makes that jump, I celebrate. That's really important to me. I also like to look at the students and what they're doing and their successes. I've got former Ph.D. students who are faculty members at eight different universities. I'm very proud of that. They've all been promoted and tenured, too. Not all, but they're making good progress, the ones that haven't been. But the successes of our students and the successes of the folks that I've mentored along the way are really what I take away. I see that as big wins. And undergraduates, too. I mean, I'm friends with lots of undergraduates, and Facebook is great, and you see different things. But to see my undergraduate students being successful in their careers and knowing I had a little part of that to play. Yeah.

I think as an active alumni, and as an active faculty member, there are lots of things I can do to continue to help the department. I think to prove our worth is big, and that's been a problem for decades, and we're still not doing that very well. We have to feed whatever the number is now, 9 billion, 10 billion people. And the people side of agriculture, the human element, is one of the biggest roles in that. How are we going to make that work? We're important in meeting that need of the world. I mean, but we're not doing a really good job of talking about that. We've always been very poor at measuring and communicating how we make a difference, and part of it is because, with humans, it's difficult. It's difficult to measure and put it into terms that everybody can understand. But we've had a lot of time to figure this out. We really still haven't done it very well. And there are ways to measure those impacts. So I think that's really important.

I think to prove our worth is big, and that's been a problem for decades, and we're still not doing that very well.

I worry, in Virginia, the next economic downturn and what might happen to Extension. I feel like Extension is really vulnerable to budget cuts at the state level. Federal budget numbers are flat. Things have gotten more expensive, and we're still trying to operate on a budget that's 20 years old. That's kind of scary. So I feel like Extension is going to come to a crossroad where they're going to have to change the system, and nobody wants to do that. It's happened in other states, and it's been successful in most states. I think we need to learn some lessons there and kind of be proactive in thinking about that in Virginia. Nobody wants to do that, though. And Agricultural Education in the secondary schools, same kinds of questions about what are we worth? The curriculum needs to be modernized.

We need to truly prepare children for entering advancement and agriculture careers. And that could mean going to college first, or going to technical school first, or coming out of high school and getting a good job. But we've got to get back to that, and what are the skills that we need to do that? It's not using an oxy-acetylene torch, which industry hasn't used for 25 years. It's other things, and so we've got to sit down and take a real look at what students need and revamp our programs to meet those needs. I think those are some real challenges.

And then internationally, there are also challenges that we can help with, too. Our expertise is very valuable internationally from youth development to teaching and learning at universities. I mean, all of those things we do in the department. But we've had projects in Senegal and in other countries where we've been able to really shine with our expertise, and I think that's going to continue to be a challenge and opportunity for us.

I think we've got a real foot in the door on the evaluation piece and proving our worth. I think that's something we can really sink our teeth into, and we should. I think we've got a real voice in modernizing curriculum for agricultural education around STEM and around new skill sets that are needed to help children enter and advance in agriculture careers, which is really what we ought to be doing in those secondary programs. I think there's a real opportunity for human and institution capacity building here and around the world. And I think there's a real opportunity to help people think about making their communities viable places to live, places where people have opportunity, places where people have the resources they need, and places where people feel like they fit and they can live comfortably. I think we've got a voice in all of those areas, and they're big opportunities for us.

I'm very proud of where the department is. I think we're really one of the top five programs in the country now, and it's been really nice to see the faculty work hard to accomplish that. And I think the next 10 years will be better than our last 10 years. So, I'm excited to see where we're going.

12

Chaney Mosley
Ph.D. 2012

Professional Roles – Middle Tennessee State University: Assistant Professor of Agricultural Education

I am an alumnus of the doctoral program. I started in the fall of 2009 and I graduated in the spring of 2012. I was affiliated with the department from 2009 to 2012. The name of the department at that time was Agricultural and Extension Education. While I was there, the department head was placed in the role of Interim Associate Dean for the College of Agriculture and Life Sciences. Because of this, we had a new Interim Department Head. I thought that was interesting. I say that, because when I first interviewed for acceptance into the department, it was with one Department Head. Then, for a year there was a different Department Head. It was just a different dynamic.

> When I started, no students had earned their doctoral degree in the Department of Agriculture and Extension Education.

I'm sure the department had an articulated mission or vision statement, but I couldn't tell you what it was exactly. But at that time, the department functioned primarily as a graduate degree granting department. We had undergraduate students that took courses from our department. Some were oral communication, written communication, and leadership. But to my knowledge, the students weren't receiving an undergraduate degree in those areas. The department also had a focus on formal and non-formal education related to leadership and Extension.

It was a very young department. What I mean by that is that the doctoral program was fairly new. In fact, I think I was in the third year that the department had doctoral students on campus. So, it was pretty young in terms of longevity. We didn't have an extensive alumni network who had completed the doctoral program. When I started, no students had earned their doctoral degree in the Department of Agriculture and Extension Education. I think that was interesting. Also, there were also young faculty. Every faculty member was an assistant professor. We had no associate professors, other than Rick, who is the department head – he was full professor. Everyone else was assistant professor. I say young faculty but I mean young in terms of their tenure – not age, but experience.

While I was there, the department was working on conducting and hosting stakeholder feedback sessions for the Virginia Cooperative Extension service. That was a statewide initiative that was a little difficult for graduate students. We were tasked with going out across the state of Virginia where meetings were being hosted. We weren't facilitating the meetings, but we were there to record the conversations, transcribe them, and code that qualitative data. It was a heavy lift, and I remember being really involved with that project during my first year. But I don't know what the outcome of that was. Another big project was the teaching lab where the department had video cameras and facial recognition software installed and was starting down the road of analyzing student expressions in reaction to the teacher. It was pretty novel. Just having that laboratory facility was unique.

More than anything, what has really impacted me is the mentorship and guidance that I got. It was invaluable and gave me a really positive experience of what a committee member could and should do.

I would say that being a graduate student and being on graduate assistantship did not influence my career, but what did influence my career was the process of earning a doctoral degree and particularly my dissertation and the topic that I researched. Even still today, I reference my dissertation. I frequently reference the methods that I used and sometimes I reference the findings. But the way that my study was structured really opened doors for me. And I think that was a product of working with my advisor, Tom Broyles. My advisor is not in the department anymore but he is here in Nashville where I live. He was an incredible advisor and continues to be a mentor and a friend to me. Eric Kaufman's guidance was important to me as well. I was able to get attention and help from both of them in organizing research and writing.

I was a graduate assistant, I did some research, and I taught classes. More than anything, what has really impacted me is the mentorship and guidance that I got. It was invaluable and gave me a really positive experience of what a committee member could and should do. I've since served on committees of other students, and I've guided students on the experience of applying to graduate school. I received a thank you note in the mail two days ago with a hardbound copy of a thesis that a masters student sent. When I opened to the acknowledgments page, I was the first person thanked. But that wouldn't have happened, and I wouldn't have had the skill set to mentor someone through that process, had it not been for the example that Tom and Eric set for me.

When I was there, I was nominated as outstanding graduate student the year I graduated. In 2015, I was named as the outstanding young alumni from the department. I have continued to engage through the College of Agriculture and Life Science's mentor program – last year I served as a mentor to an undergraduate student. I've tried to stay engaged when I can. And also, I've worked with the CALS alumni directory to facilitate and coordinate some events here in Nashville, because the College coordinates regional events. As far as the department is concerned, I get random phone calls, emails, and text messages from people. So, I think that the relationship is still there, but because I'm not in the state, I'm not really able to actively engage.

I think there are opportunities to interact with the department, but they depend on the career. There are certainly opportunities to collaborate with faculty members on research projects. I already know the intersections of my research interests and those of the faculty, and I already have those working relationships, so I think that's helpful. I think the value of a graduate degree, specifically a doctoral degree from our department, is often different from other colleges and departments that offer a doctoral degree in agricultural education or a related field. I think that in many universities, students will perceive their Ph.D. and their career track is academia. Rick and I talked about the value of looking back at our doctoral alumni and what they have done and are doing with their doctoral degree outside of academia. Because, based on my knowledge, the majority of us who have graduated are not in academia and are not working in higher education. We've taken the knowledge and skills and applied it in related contexts, but I think that those examples would be great to show students their career options outside of academia. You asked how I'd like to be involved, and I think that it would be interesting to conduct a case study around students who earn doctoral degrees from our department, what they're doing now, with a focus on income and earning potential. The connections that

I made in grad school have been really valuable to my current role, and I know that many of us are doing amazing work. And because we've chosen to not enter academia, I think our salaries are often higher than they would be had we gone on the academia track.

The biggest challenge is... how to feed a growing population with a shrinking land mass.

One of the challenges facing agricultural education and extension is image. And that's not only agricultural education, but all of career and technical education. But we certainly have a challenge related to image, because many think of formal agricultural education as vocational education. I think the biggest challenge with extension education is awareness. I'm not sure that the community, by in large, is aware of the resources that the Extension service can offer them. Another challenge with agricultural education is consumer awareness and education. Agricultural educators shoulder the weight of that challenge to communicate about the industry to the public through extension and formal education, and I don't see that going away. There is a question on how to produce future agriculturalists—those who will produce the food, those who will innovate the food system, and those who will innovate technology required to enhance the efficiency of our food system. The answer is through agricultural extension and education. And with that in mind, the biggest challenge is—which faces the entire agricultural industry—how to feed a growing population with a shrinking land mass.

Let me focus on agricultural education: here is a huge opportunity in workforce development. We have experienced an increase of career opportunities in the field that is vaguely referred to as "agricultural technology" or "agricultural customer service", but what it really boils down to is agricultural education. Ag tech is huge; there are a lot of IT companies that are focused on agricultural products. Meaning, their target customer is a producer, and they're creating IT based programs that will help a farmer be more efficient. Those programs provide weather patterns and schedules on when to apply nitrogen, for example. The problem is that when producers subscribe to these programs, they may not experience success using them. So many of the IT companies are trying to hire people to serve the role of customer success—and the role can be called a lot of different things. Essentially, it is a client management role where you go out and you work with the person who's purchased that product and train them on how to use the product. You're in constant communication with the client and build a relationship with them to show them the value of their product and make sure they're using it correctly. And after a year or two at the end of their subscription, they want to continue using the product. But without that key person, often the subscriptions may not be renewed or valued. That is where ag education comes in. I think for years we've thought of the career trajectory for ag education majors to be either be a school-based agriculture teacher or an extension Agent. There are other opportunities in between, but what I just described requires the ability to plan to teach someone something and to assess whether or not that person understands and can do what we've taught them. And that's an ag teacher. I think ag and extension education majors can be siloed into thinking that their career options are limited, but no, there are incredible opportunities. I would even rope data science into that, because teachers and extension agents are constantly evaluating their programs and students, collecting and analyzing data. We have what these companies need, and we can fill the need that is currently in the ag tech workforce.

13

Kerry Priest

Ph.D., 2012

Professional Roles - Kansas State University : Staley School of Leadership Studies, Associate Professor

I graduated with my Ph.D. in 2012 from the Agricultural and Extension Education department with an emphasis in collaborative community leadership and leadership education. When I came in 2008, the doctoral program was still fairly new; I was part of an early cohort of students. And it was exciting to be part of a new and growing program. My primary responsibility as a graduate assistant was working with the Residential Leadership Community (RLC).

> It was exciting to have a role in shaping the vision, strategic plan, and the curriculum for the RLC.

When I came to Virginia Tech, the RLC had just merged with the Ag and Extension Education department. One of my first responsibilities was helping physically move the office. I worked for several years as a graduate assistant, responsible for teaching, curriculum development, and programming. In the final year of my Ph.D. program, I served as the interim director for the RLC. It was exciting to have a role in shaping the vision, strategic plan, and the curriculum for the RLC. During that time, we really saw a stronger connection and greater collaboration between academic affairs and the student affairs, as well as a more intentional integration of service learning. It was, of course, difficult to leave a program in which I had learned and invested so much. But it really prepared me for my next step, which was moving into an assistant professor of leadership education at K-State.

During my time at Virginia Tech, we experienced a lot of growth and change. We had a lot of new students coming in and new faculty joining us, so there was an opportunity to really learn from a diverse cohort and group of instructors during that time. I've been gone from Virginia Tech now longer than I was there, but it still feels like it was yesterday – it was such a formative time.

As I think about the purpose of the department, I recall being a part of negotiating our name change from Agriculture and Extension Education to Agricultural, Leadership, and Community Education. We had a lot of conversations about how our name reflects who we are, what we do, and who our students are becoming. On one hand, we very much focused on teacher preparation – developing future agriculture education teachers and extension agents. Additionally, we were building the community education piece in terms of understanding how change happens within communities, especially through grassroots efforts. There was an emphasis on developing beginning farmers. At the same time, we had a strong focus on undergraduate leadership education through the Leadership and Social Change minor. Our graduate program also focused on preparing people to work in leadership education and development roles. So it was a very complex set of connected purposes that all centered around preparing people to do work that advances change.

> I got a great education through my courses, but then I was able to exercise leadership within the RLC program.

In terms of accomplishments, I think our department has a strong emphasis on scholarship, with many faculty and graduate students producing journal articles and other publications, presenting at conferences, and really building the reputation for AEE (ALCE) at Virginia Tech as a quality program. I also feel like the RLC was a model learning community, integrating VT's Aspirations for Student Learning through curricular and co-curricular work. I recall building more programming and recognition for our students around those aspirations, and so we were able to be more connected to the institution in that way. Many of our students were just doing really excellent work, and we were able to acknowledge and recognize some of those students for their service and leadership, not just within the department but campus-wide. Finally, as a young Ph.D. program, we were graduating students who were able to go out and be productive in roles both inside and outside higher education, and inside and outside of agriculture.

I had an amazing hands-on learning experience as a graduate assistant, teaching undergraduate leadership education courses. I was able to have ownership of a classroom, be involved in curriculum development and evaluation, and also the project management side of a leadership minor program, including the use of service learning as a high impact practice. And at the same time, my graduate studies in leadership were complementing and supporting this work. So I got a great education through my courses, but then I was able to exercise leadership within the RLC program. So when it came time for me to apply for a tenure track role in an interdisciplinary leadership education program, I was able to leverage that experience. I had already been doing the work of someone in that role, and I really, really appreciated that.

My classes were also beneficial in helping me determine a scholarly trajectory. I am thankful for the opportunity I had to attend conferences, as it helped me get connected to professional networks. So right away I was connected to communities of practice that have really supported me and provided an opportunity for professional service. Part of being on a tenure track is you have to build your reputation as a scholar. And I'm thankful for the opportunities at Tech that set me up to be successful in that later on. Going through the dissertation process, I had great mentors and supporters. Now as I'm preparing to mentor doctoral students myself, I have good models to draw from.

I just earned tenure and promotion to Associate Professor. Looking back, I realize that my path to that did not just start when I went to K-State. The last decade of my life, including my Ph.D., were preparing me for this moment and for what's next.

> I'm really committed to the idea that, through our work, we can also make communities better.

As a leadership educator, I feel like my work has an impact in the development of students. And I care deeply about my students and helping them think about their lives and their roles and what they're trying to become and how they can apply, through leadership, apply to their majors or their technical skills, how they see themselves as having a role in creating change in the world.

I think through my scholarship, too, around teaching and learning of leadership and around identity work, and even identity work of educators, I feel like I've been able to make a contribution to the field and support how do we do this work, and helping others become and belong within our community of practice as well. And also thinking about just the civic purposes of our work. This is a land grant. I'm at a land grant institution. And recognizing that we're not just developing skilled workers, right, we're developing citizens. And how do we think about the public purposes of our work and, how do we, through community engagement and community engagement scholarship, not just develop individuals, but also develop the communities that we're a part of? And so I'm really committed to the idea that, through our work, we can also make communities better, and trying to explore that in my own classrooms and through scholarship. And that's exciting to me. I like the idea of having that purpose, that public purpose.

I try to stay connected with Virginia Tech faculty and students through conferences and following what's happening on Facebook. The opportunity to come back and visit reminds me how much I loved being here at Virginia Tech and all the great relationships I still have. I would like to stay involved as an alumni, for example connecting with classes or supporting graduate students. There may be ways to continue to partner on projects as we recognize common interests. And I think it is important to continue to support each other as colleagues.

While my background and education was situated in a college of agriculture, my current role is interdisciplinary. One thing I think that impacts all of us who are committed to education (leadership, agriculture, community) are questions of social justice. This means not just integrating ideas of diversity, but also practicing inclusion, equity, and justice into the work that we're doing. We live in really polarizing times, with many tough challenges facing our world – social, political, economic, environmental. And the way we can make progress on those challenges isn't just more technical skills – which do matter. But, we also have to be able to work with people who are very different than us. What I'm thinking a lot about now is, how do we truly integrate inter-cultural development and multicultural learning into our work? This includes intentionally trying to dismantle systems of oppression that exist within our institutions and in our lives. And that's a really hard thing to talk about, because it feels political and polarized. But we have to do this work, because what's happening now is not working.

I'm really interested in this idea of vertical development that comes from the Center for Creative Leadership. It's the idea "filling your cup" with technical skills and competencies versus "growing your cup" by building our capacity to recognize and engage in an interconnected world. How do we deal with ambiguity and uncertainty and not relying on authority to tell us what to do? How do we develop those capacities to be interdependent? This is the lens we need to make progress. The way that we develop those capacities has to be through engaging multiple perspectives. Putting ourselves in situations that we'd call "high heat," – which are uncomfortable, ambiguous, and probably involve failure. And then we have to engage in deep, intense meaning-making through mentoring, advising, and reflection. And so to me, that gives insight into the role of higher education – we need to create the conditions for learning that provide those experiences. And so, how do we do that? It's difficult, because our system is set up in a way that often just keeps replicating knowledge transfer. So I think authentic engagement, problem-solving, developing critical-thinking, and intentional inter-cultural development is absolutely critical – in the field of Agriculture and for all of us!

I think authentic engagement, problem-solving, developing critical-thinking, and intentional inter-cultural development is absolutely critical

Someone asked me a couple weeks ago, "What do you think the future of higher education looks like?" And I was like, "I have no idea. But I know it's not going to look like it does now," right? So I think we have to assume that the methods and the processes and even the knowledge that we have now is not going to sustain us for the next 20 years. We're going to have to be flexible and comfortable with ambiguity and willing to experiment in lots of different ways with different modes of doing things. In some ways, the field of agriculture has always been that way. Although I recognize there's a lot of our systems that are kind of slow to change, people who are engaged in agriculture are used to uncertainty. It's a volatile and risky business. How do we learn to navigate the complexities even better, and tap into the context that we're used to? How do we prepare to navigate those changes even more nimbly?

This project itself is a really great opportunity to learn from many perspectives. So if you think about it, what you're doing here is a way of exercising leadership for change! That is exciting.

14

Rebekah Slabach

B.S. 2016

Professional Role – Virginia Cooperative Extension: County Extension Agent, Agriculture & Natural Resources, Halifax County

I was a student here from 2011 to 2016. I graduated with a degree in ag sciences and also had a leadership and social change minor. When I started, it was the Department of Agriculture and Extension Education, and when I graduated it had been changed to ALCE in '16.

So, significant change number one – the Residential Leadership Community (RLC) has really grown, especially with the leadership and social change minor gathering more interests for students outside the department in college. And then also, of course, the restructuring from AEE to ALCE. So adding more courses specifically with education, extension, and also creating student research opportunities for independent studies. The primary purpose of the department is to equip students to be leaders and engage with whatever organization they were going to be involved with. So, probably working with youth or producers, or community.

The primary purpose of the department is to equip students to be leaders and engage with whatever organization they were going to be involved with.

Some of the challenges that the department had to overcome were adapting to the new needs and new interests. When the RLC first started, it just was for ag students primarily, and trying to grow that. And how to restructure that to be interdisciplinary, and have experiential learning opportunities, and service projects. And then also, reconfiguring the department's identity with, how do we reach and prepare students to not only be teachers, but also Extension Agents, or also community educators.

Accomplishments during my time in the department were creating more opportunities for students – whether with internships, independent studies, research opportunities, encouraging study abroad, partnerships with professors and undergrad students. The RLC, again, just the way different classes were structured where students had to interview industry professionals or be connected with a local school or community. So challenging that learning continuum.

I'm serving as an agriculture extension agent. So the department and my education not only gave me knowledge and understanding of extension; that perspective of how land-grant universities work and the research component and the outreach aspect. But also the soft skills that critical thinking, problem-solving aspect through internships for research and independent studies. So every day, I'm influencing the local food sector of my community. Whether it's helping producers, getting them connected with other aspects of the food chain. Or the farmers' markets connecting local citizens with local food. So, helping our producers be productive, but also helping consumers be connected and understanding their food source.

So every day, I'm influencing the local food sector of my community.

I just graduated two years ago, so I'm hoping to serve the department and help in any way I can in the future, as a proud alumni. I've come back with Alpha Zeta, which has some advisors and students in the department, and have been there for some of their alumni things. And I helped serve as a contact. I was an extension agent interviewee for the Introduction to Cooperative Extension class. So I hope to see more in the future with my role as Extension agent and see how to be connected with what the department needs. I would like to encourage students to make the most of opportunities that they can, whether it's sharing things I've been a part of or getting them connected with Extension internships or experiences or a class project with a Halifax issue. Anything that might connect them with what I'm working on.

So, greatest challenges facing agricultural education and Extension – education-wise, we're in a period of transition, I think, with shifts with career and technical education requirements and standards, and what that looks like for funding and teachers. So how to equip ag teachers to go in, but also to be able to reach and start new chapters and also urban programs. It's going to be a new challenge. And then, Extension-wise, helping bring young people in and help them be connected and how to be able to reach a new generation. The millennials, Generation X, completely different than older generations with their interest in their food supply and knowledge. And so being able to help overcome misconceptions and perceptions about agriculture effectively is also going to be important.

Opportunities are very wide open. Again, I think starting at the collegiate level, I think if we can get young people involved and connected, we can use worldwide problems as projects or case studies for them to work on get involved with. And we can help them get plugged in on either a local or a regional or a global level in different ways to get them connected. And help all of that be better.

This department gives you great knowledge and perspective, but also gives you opportunities to deepen your soft skills. So it's important to have that knowledge, but to know how to work with people, how to have that global agriculture perspective, how to understand problems, how to solve them critically, how to connect the dots. That's also important. And making the most of opportunities helps make that full force.

Made in United States
North Haven, CT
04 November 2026

21771558R00053